ZOONOSES

Infectious Diseases of Animals
Transmissible to Humans

The Author

Dr. Chandra Shekhar is currently Associate Professor & Head, Department of Veterinary Public Health & Epidemiology, College of Veterinary Science & Animal Husbandry, Narendra Deva University of Agriculture & Technology, Kumarganj, Faizabad (U.P.). He obtained his B.V. Sc. & A.H. and M.V.Sc. (Veterinary Public Health) degrees from College of Veterinary Science & Animal Husbandry, Mathura, C.S.A.U.A&T, Kanpur. He earned his Ph.D. from College of Veterinary & Animal Sciences, G.B.P.U.A.&T., Pantnagar. He has reported first *Salmonella* Drogana serovar from human source in India. He has to his credit more than 50 scientific papers in various journals of repute. Dr. Chandra Shekhar has also attended and presented papers in various national and international conferences/seminars. He is a life member of *Indian Association of Veterinary Public Health Specialists* (IAVPHS) and member of the editorial board of *"The Asian Journal of Animal Science"*, *"Veterinary Science Research Journal"* and *"Research Journal of Animal Husbandry and Dairy Science"*. He is the reviewer of the journal *"International Journal of Basic and Applied Agricultural Research"*. He has been conferred an award the "Fellow" by *"Hind Agri-Horticultural Society (FHAS)"* Muzaffarnagar (U.P.), India. Besides this he has been the recipient of many honors, awards and distinctions from various scientific bodies.

ZOONOSES

Infectious Diseases of Animals Transmissible to Humans

– Author –

Chandra Shekhar

M. V. Sc., Ph. D.

Associate Professor & Head
Department of Veterinary Public Health & Epidemiology
College of Veterinary Science & Animal Husbandry
N.D. University of Agriculture & Technology
Kumarganj – 224 229, Faizabad, U.P., India

2018

Daya Publishing House®

A Division of

Astral International Pvt. Ltd.

New Delhi – 110 002

ISBN 9789387057890 (International Edition)

Publisher's Note:

Published by   :   **Daya Publishing House®**
                                      A Division of
                                      Astral International Pvt. Ltd.
                                      – ISO 9001:2015 Certified Company –
                                      4736/23, Ansari Road, Darya Ganj
                                      New Delhi-110 002
                                      Ph. 011-43549197, 23278134
                                      E-mail: info@astralint.com
                                      Website: www.astralint.com

Dedication

THIS BOOK IS DEDICATED TO STUDENTS/READERS

Foreword

This book on "Zoonoses: Infectious Diseases of Animals Transmissible to Humans" authored by Dr. Chandra Shekhar, Associate Professor & Head, Department of Veterinary Public Health & Epidemiology, College of Veterinary Science & Animal Husbandry, Narendra Deva University of Agriculture & Technology, Kumarganj, Faizabad (U. P.) has been successfully planned to fulfill the long-standing need for a textbook on Zoonoses for the undergraduate students of Veterinary and Animal Sciences. This book can also be useful for Veterinary Public Health professionals.

In spite of significant progress made in the field of public health services, the number of zoonotic diseases are continuously increasing and affecting the population. This increase has been attributed to many factors such as, global warming, changes in farming practices, human demographics and behavior, microbial adaptation, technological advancement, industrialization, social and cultural factors, breakdown of host's defense mechanisms as well as international travel and trade. Considering these factors into account, the author has made an effort to disseminate the information on zoonoses through this book to the veterinary students, scientists, teachers, veterinary professionals including the professionals in dairy and meat industries.

This book has been divided into 50 chapters and 4 appendices in which each aspect has been systematically described. The presentation of the book is simple, straightforward and lucid. The author has put his more than 13 years of experience as a teacher in Veterinary Public Health to make this book informative. The author deserves appreciation for this sincere effort. I am confident that this book may attract attention of the students, teachers and scientists and those who are associated with the Veterinary Public Health services in the country and elsewhere. I wish him great success for this noble work.

Dr. S. P. Singh

Professor & Head
Department of Veterinary Public Health & Epidemiology,
C.V.A.Sc., G.B. Pant University of Agiculture & Technology,
Pantnagar

Preface

The essential role of animals in the transmission of infectious diseases has long been recognized. Animals maintain the infections in nature by harbouring and enabling pathogens to survive and active cause of pathogen spread in the environment and to other animals, including humans. Zoonotic diseases have a significant effect on public health worldwide, showing a higher incidence rate in developing countries due to inadequate diagnostic laboratory services, lack of control strategies, as well as the lack of education and awareness in the communities. Therefore, collaboration between medical and veterinary scientists as well as public health practitioners and laboratory scientists is essential in order to investigate new and emerging zoonotic diseases. Fully equipped laboratories containing subtyping techniques are also essential for detecting disease outbreaks and characterizing transmission routes. By using molecular subtyping, different strains can be differentiated on the basis of their genotypes and phenotypes. In addition to laboratory techniques, the work of epidemiologists, ecologists, and environmentalists is crucial in establishing patterns and preventing outbreaks of zoonotic diseases. Moreover, health education to the public can help to prevent and control the spread of zoonotic diseases. If public know about causative agent and its reservoir (*e.g.*, association of dog with rabies, pigs with Japanese encephalitis and monkeys with Kyasanur forest disease), they can be more careful about zoonotic diseases and thus they can prevent the spread of the diseases.

There has been a persistent demand from the veterinary students to write a book on zoonoses that can provide complete information. The effort has been made to provide current scientific information in the book. In this book, all sections of zoonoses that is general aspects of zoonoses and zoonoses caused by different types of pathogens has been describes in details. Most of the important zoonotic diseases have been presented in the book. In addition, a large number of zoonotic diseases have been briefly presented in a table. The main feature of the book is that the diseases and their management in animals and humans have been separately described. Other features of the book include transmission of diseases that has been clearly explained with figures which will make easy to understand the infection cycles of zoonotic diseases. Appendices of the book include glossary of terms, zoonoses involving various animal species, infections resulting from animal bite and principal features of important zoonotic diseases. I hope the book will be very informative and useful for B. V. Sc. & A. H. students and also for postgraduate students of Veterinary Public Health and veterinary professionals.

I am highly indebted to Dr. S. P. Singh, Professor & Head, and Dr. A. K. Upadhyay, Professor, Department of Veterinary Public Health & Epidemiology, College of Veterinary & Animal Sciences, G.B. Pant University of Agriculture & Technology, Pantnagar (U.S. Nagar) Uttarakhand for their blessings, invaluable suggestions, constant inspiration, encouragement and support, so that I could write this book. I express my heartiest gratitude to Dr. Ashok Kumar, ADG (Animal Husbandry), ICAR, Pusa, New Delhi; Dr. R.K. Agrawal, Principal Scientist; Dr. S. V. S. Malik, Principal Scientist & Head and Dr. Z. B. Dubal, Senior Scientist, Division of Veterinary Public Health, ICAR-Indian Veterinary Research Institute, Izatnagar, Bareilly for the encouragement, support and inspiration they provided to me. I am grateful to Dr. Basanti Bisht (former Professor & Head) and Dr. Udit Jain, Assistant Professor, Department of Veterinary Public Health, U. P. Pt. Deen Dayal Upadhyay Pashu Chikitsa Vigyan Vishwavidyalaya Evam Go Anusandhan Sansthan, Mathura for their support and encouragement. I thank to my all students, friends and well-wishers for their continued support in this endeavour. I am highly thankful to Dr. B.B. Singh (Editor), Astral International Pvt. Ltd. for his nice cooperation and keen interest in publication of this book.

Chandra Shekhar

Contents

Section I: General Aspects of Zoonoses

Section II: Bacterial Zoonoses

Section III: Viral Zoonoses

Section IV: Fungal Zoonoses

Section V: Parasitic Zoonoses

Appendices

Abbreviations Used in the Book

@:	Commercial at or at
°C:	Degree Celsius
AIDS:	Acquired immunodeficiency syndrome
BCG:	Bacille Calmette Guerin
BSL:	Bio-safety level
CAT:	Computed axial tomography
CDC:	Centers for Disease Control and Prevention
CFT:	Complement-fixation test
cm:	Centimetre
CNS:	Central nervous system
CSF:	Cerebrospinal fluid
CT:	Computed tomography
DNA:	Deoxyribonucleic acid
e.g.:	*Exempli gratia* (for example)
EDTA:	Ethylene diamine tetra acetic acid
EIA:	Enzyme immunoassay
ELISA:	Enzyme-linked immunosorbent assay
FAO:	Food and Agriculture Organization
FAT:	Fluorescent antibody technique
FFP:	Full face protection
HIV:	Human immunodeficiency virus

i.m.: Intramuscular

i.v.: Intravenous

ICMR: Indian Council of Medical Research

IFA: Immunofluorescence assay

IFT: Indirect fluorescent antibody test

IHA: Indirect haemagglutination

Ig: Immunoglobulin

IHR: International Health Regulations

INFOSAN: International Food Safety Authorities Network

INH: Hydrazide of isonicotinic acid

IU: International unit

Kg: Kilogram

L: Litre

LASER: Light Amplification by Stimulated Emission of Radiation

m: Metre

mg: Milligram

ml: Millilitre

MRC 5 cells: Medical Research Council cell strain 5

MRI: Magnetic resonance imaging

NNN agar: Novy-McNeal-Nicolle agar

OIE: Office International des Epizooties

ORS: Oral rehydration salts

PCR: Polymerase chain reaction

PFGE: Pulse-field gel electrophoresis

PPE: Personal protective equipment

RBT: Rose Bengal test

RFLP: Restriction fragment length polymorphism

RIA: Radioimmunoassay

RNA: Ribonucleic acid

r-RNA: Ribosomal RNA

spp.: Species

Viz.: Videlicet (namely)

WHO: World Health Organization

Wt.: Weight

α: Alpha

Section-I

General Aspects of Zoonoses

The role of animals in the transmission of infectious diseases has long been recognised. In general, zoonotic pathogens can survive and multiply in the animals or environment without the presence of human. In addition, animals often provide a medium for the survival and reproduction of arthropod vectors which are responsible for transmission of many zoonotic pathogens. Human acts as dead-end host in rabies, brucellosis, trichinellosis *etc.*, therefore, human is not essential for the perpetuation of the pathogenic agent. The link between human and animals with their surrounding are very close especially in developing countries which can lead to serious risk to the public health with severe economic consequences. Wild carnivores and stray dogs are major reservoir of rabies in India. Pig on the other hand also known as "Mixing Vessels" for influenza viruses where, reassortment of various genes of influenza viruses easily takes place. Pig also acts as an amplifying host for Japanese encephalitis. As far as birds are concerned, the entry of migratory birds in wildlife sanctuaries as well as water bodies could not be stopped, thus playing a vital role in transmission of avian influenza. The role of rodents in transmission of zoonoses has also been accepted worldwide. India is facing problem of *Hantavirus* infection, Kyasanur forest disease and Crimean Congo hemorrhagic fever which are recently reported as rodent-borne zoonotic infections. Animals play important role in the perpetuation and spread of most of the zoonotic pathogens, while humans can play important role in perpetuation and spread of some zoonotic pathogens like *Taenia solium* and *Taenia saginata,* for which humans are the unique definitive host. Human may also be a definitive host of *Diphyllobothrium latum, Paragonimus* spp., *Sarcocystis* spp. and some schistosomes and very important host in American trypanosomiasis. There are many factors for increased occurrence of zoonotic diseases including deficiency in public health infrastructure, high population density, alteration of the environment (changes in ecology and climate), microbial adaptation and change (evolution of new strains of microbes), establishment of human settlements in formerly uninhabited areas, intensification of animal production and increase in the trade of live animals, animal products and other foodstuffs.

Chapter 1

Definition, Objectives and Socio-economic Consequences of Zoonoses

Zoonoses

The origin of the word zoonoses could be traced back to Greek words "zoon" meaning animal and "nosos" meaning disease. The term zoonosis was first used by Rudolf Virchow in 1855 in his famous "Handbook of Communicable Diseases" to describe the animal diseases secondarily transmissible to humans. Zoonosis is any disease or infection that is naturally transmissible from vertebrate animals to humans. Animals thus play an essential role in maintaining zoonotic infections in nature. Many of the major zoonotic diseases cause public health problem and prevent the efficient production of foods of animal origin and create obstacles to international trade in animal products.

The WHO Expert Committee (1959) defined the zoonoses as "Those diseases and infections that are naturally transmitted between vertebrate animals and man". All true zoonoses conform to the requirement of this definition-

1. The transmission of disease occurs under natural conditions.
2. The disease can also be transmitted from human to animals.
3. The toxicities and incidental infections are also considered forms of zoonotic affections.

If an ectoparasite burrows into or successfully enters the body of its host, the disease caused by it can also be considered as zoonosis. Some experts believe that zoonoses should include all animal related human health problems like animal bites. High densities of animal and human populations, close interactions between animal and man especially in rural areas, poor infrastructure of health care and diagnostic services, unavailability of adequate financial assistance and tropical climate (favours growth of microbes and vectors) provide ideal conditions for efficient and uninterrupted transmission of zoonotic diseases in the country. In developing countries there is little awareness about transmission and risks of zoonotic diseases. Huge animal population in the developing countries also favours the easy spread of zoonotic diseases. The WHO coordinates global programmes on zoonoses which include diarrhea, parasitic and communicable diseases. Other related activities are concern with vector biology and control, diagnostics and environmental health. The publications of the WHO are of importance in dissemination of information about global trends of zoonotic diseases.

Euzoonosis

Euzoonosis is an obligatory association where human is a final host and the animal intermediate host for the same parasite, *e.g.*, obligatory cyclozoonosis (taeniasis).

True Zoonosis

Zoonosis that efficiently transmitted between animals and human is known as true zoonosis.

Emerging Zoonoses

Neslin (1992) defined emerging zoonoses as "those zoonotic diseases caused either by apparently new agents or by previously known microorganisms, appearing in places or species in which the disease was previously unknown". Ledurburg (1992) defined emerging zoonoses as "those diseases whose occurrence has increased in past two decades or threat to increase in near future". WHO/FAO/OIE (2004) joint consultation on emerging zoonotic diseases held in Geneva from 3-5 May, 2004, defined the emerging zoonosis as "a zoonosis that is newly recognized or newly evolved or that has occurred previously but shows an increase in incidence or expansion in geographical, host or vector range", *e.g.*, avian influenza, severe acute respiratory syndrome (SARS), bovine spongiform encephalopathy, Nipah virus infection, *Hantavirus* infections, salmonellosis, campylobacteriosis, enterohemorrhagic *E. coli* infection *etc.* The important factors responsible for emergence of zoonoses are follows:

1. Environmental Change (Ecological Change)

Ecological changes due to agricultural and economic development are among the most frequently identified factors in the emergence of zoonoses. Accelerated degradation of the natural environment, notably in developed countries by deforestation, building of dams and land consolidation may cause wildlife species to move to new areas, favouring their relocation in sub-urban zones, therefore entering into contact with humans. Ecological factors usually precipitate emergence by placing people in contact with a natural reservoir or host for an infection hitherto unfamiliar but usually already present (often a zoonotic or arthropod-borne infection), either by increasing proximity or, often, also by changing conditions so as to favour an increased population of the microbe or its natural host. *Hantavirus* pulmonary syndrome (HPS) emerged in southern USA in 1993 due to weather anomalies. It is likely that the virus has long been present in mouse populations but an unusually mild and wet winter and spring in that area led to an increased rodent population in the spring and summer and thus to greater opportunities for people to come in contact with infected rodents (and, hence, with the virus); it has been suggested that the weather anomaly was due to large-scale climatic effects. The emergence of Lyme disease in the United States and Europe was probably due to reforestation, which increased the population of deer and the deer tick, and the vector of Lyme disease. The movement of people into these areas placed a larger population in close proximity to the vector. In India, the deforestation in Karnataka state and grazing of cattle in this forested areas led to emergence of Kyasanur forest disease.

2. Pathogen Change (Microbial Adaptation and Change)

Microbial evolution and response to selection in the environment are the major factors for emergence of many zoonotic diseases, *e.g.*, antigenic drift in influenza virus and possibly genetic changes in severe acute respiratory syndrome (SARS) *Coronavirus* in humans, development of antimicrobial resistance in numerous bacterial spp. (*e.g.*, multi-drug resistance in *Mycobacterium* spp. and *Salmonella* spp.) and chloroquene resistance in malaria parasite.

3. Change in Farming Practices

Use of pesticides leads to the development of resistance in many vectors and results into spread of many diseases. Use of excess fertilizers in the rice field leads to propagation of mosquitoes and leads to spread of Japanese encephalitis. Irrigation is one of the agricultural practices used extensively for increased production and frequently associated with disease emergence. This may be associated with the emergence of mosquito-borne diseases and snail infestation. Infections transmitted by mosquitoes or other arthropods are often stimulated by expansion of standing water.

4. Changes in Human Demographics and Behaviour

Factors in human behaviour such as commercial sex trade, intravenous drug use, outdoor recreation, use of child care facilities and other high density settings are responsible for emergence of some zoonotic diseases, *e.g.*, HIV and other sexually transmitted diseases and spread of dengue (mainly due to urbanization). Rural urbanization allows infections arising in isolated rural areas, which may once have remained obscure and localized, to reach larger populations. Increased number of HIV cases in Asia is mainly due to this dynamic. Human behaviour can have important effects on disease dissemination. The best known examples are sexually transmitted diseases, and the ways in which such human behaviour as sex or intravenous drug use have contributed to the emergence of HIV are now well known.

5. International Travel and Trade

Worldwide movement of goods and people and air travel leads to the emergence of some diseases, *e.g.*, dissemination of HIV, dissemination of mosquito vectors (*Aedes albopictus*-Asian tiger mosquito), rat-borne Hantaviruses, introduction of cholera in South America and dissemination of *Vibrio cholerae* O139 via ships.

6. Technology and Industry

Food production and processing, globalization of food supply and change in food processing and packaging are the major factors for emergence of many important zoonotic diseases, *e.g.*, enterohemorrhagic *E. coli* (EHEC), bovine spongiform encephalopathy (BSE), Creutzfeldt Jacob disease (CJD), SARS, avian influenza and Nipah virus infection. The industrialization provides the opportunity to introduce agents from far away. A pathogen present in some of the raw material may find its way into a large batch of final product, as happened with the contamination of hamburger meat by *E. coli* strains causing hemolytic uremic syndrome. Bovine spongiform encephalopathy (BSE), which emerged in Britain, was likely an interspecies transfer of scrapie from sheep to cattle that occurred when changes in rendering processes led to incomplete inactivation of scrapie agent in sheep byproducts fed to cattle.

7. Deficiency in Public Health Infrastructures

The rapid spread of cholera in South America may have been abetted by reductions in chlorine levels used to treat water supplies. These problems are more severe in developing countries. The US outbreak of waterborne *Cryptosporidium* infection in Milwaukee, Wisconsin, in the spring of 1993, with over 400,000 estimated cases, was in part due to a non-functioning water filtration plant.

8. Breakdown in Public Health or Control Measures

Curtailment or reduction in the prevention programmes and lack of or inadequate sanitation and vector control measures are important factors for emergence of some zoonotic diseases, *e.g.*, emergence of TB in USA, cholera in refugee camp in Africa and diphtheria in former Soviet Republic and Eastern Europe in 1990s.

9. Breakdown of Host's Defense

An immune depression following either medication or infection caused by pathogenic agents capable of weakening the host's immune defences allowing infection by opportunistic organism can result from the breakdown of the host's defences. Immunosuppression results the emergence of *Mycobacterium bovis* and *Listeria monocytogenes* in humans. Immunodeficiency results the development of HIV infection.

10. Social and Cultural Factors

Social and cultural factors such as food habits and religious beliefs also play important role in the emergence of zoonoses.

Re-emerging Zoonoses

Re-emerging zoonoses may be defined as "If the disease had been present at the location in the past and was considered eradicated or controlled, the disease is considered to be re-emerging". Re-emerging zoonoses may also be defined as "the reappearance of zoonotic disease after a gap of long period, in a geographical area where the disease was completely eliminated/eradicated". The OIE has defined the re-emerging disease as "an already known disease that either shifts its geographical settings or expands its host range or significantly increases its prevalence".

The contact with wildlife during hunting, fishing or ecotourism has led to the re-emergence of organisms such as *Leptospira* spp., *Bartonella* spp. and *Francisella tularensis* are excreted by healthy animals. Plague is re-emerging disease in Madagascar. Malaria is re-emerging in Africa (malaria prevalence is increased due to resistance developed to the drugs used to kill the parasites and mosquito vectors). Chikungunya fever is re-emerging disease in Asia. Cholera has increased incidence due to increase in shipping. Avian influenza, West Nile virus infection, bovine tuberculosis in wildlife, Lyme disease and leptospirosis are some other examples of re-emerging zoonotic diseases. Re-emergence of disease is mainly due to international travel and trade.

Lingering Zoonoses

Some of the "lingering" zoonoses are re-emerging in some regions, although they seem to attract less public awareness, *e.g.*, brucellosis, dog rabies and parasitic diseases such as cysticercosis/taeniasis and echinococcosis/hydatidosis.

Neglected Zoonoses

Except for the newly emerging zoonoses such as SARS and highly pathogenic avian influenza H5N1, the vast majorities are not prioritized by health systems at national and international levels and are therefore labeled as neglected. They affect hundreds of thousands of people especially in developing countries, although most of them can be prevented.

Occupational Zoonoses

Specialized professional activities provide opportunities for intimate contact between man and animals or animal products. These activities enhance the exposure to the infectious agents and risk factors for transmission of zoonotic diseases. Schwabe has listed seven professional groups associated with occupational zoonoses such as agriculture, animal product manufacture, laboratory, epidemiological and clinical, recreational, sylvan and campestral and emergency. Each of these groups includes several professional activities *e.g.*, farmers, veterinarians and animal handlers are part of agriculture activities. Physicians, nurses, health professionals and veterinarian constitute epidemiolological and clinical laboratory workers groups.

Objectives of Zoonoses

The important objectives of zoonoses are given as follows:

1. To prevent, control and eradicate the zoonotic diseases.
2. To prevent economic losses and health hazard in human beings due to zoonotic diseases.

3. To induce effective inspection of foods (ante-mortem inspection of animals, post-mortem inspection of carcase and offal and microbiological examination of milk and milk products and other foods) in order to minimize the risks of zoonotic diseases.

4. To maintain the hygienic conditions of slaughterhouses, dairy farms and other food processing plants to avoid the risks of zoonotic diseases.

5. To educate and motivate the people about source of infection, mode of transmission, personal hygiene, environmental sanitation and control measures against commonly occurring zoonotic diseases.

6. To provide opportunities for a close collaboration between veterinarians and medical professionals in their effective domains of activities. The veterinarians and physicians contribute effectively to eliminate the threat of zoonotic diseases. The effective collaboration between veterinarians and physicians can help in various ways:

    i. Assist in preventing the transmission of zoonotic diseases between humans and animals.

    ii. Assist in the implementation of policies related to control and eradication of zoonotic diseases.

    iii. Help in the surveillance of zoonoses in animals and humans.

    iv. Help in organizing the programmes for manpower training and research.

    v. Facilitate epidemiological investigation and diagnosis of zoonoses in humans and animals.

    vi. Promote studies on comparative pathology of sporadic, chronic and degenerative diseases.

Socio-economic Consequences of Zoonoses

Economic impact of zoonotic diseases can be estimated in animals as well as in human beings. At present, there are over 200 zoonotic diseases. In developing countries including India, currently there is no systematic data on zoonotic diseases is available to estimate the socio-economic impact of zoonotic diseases. Plague killed 12 million people since 1898. Rabies causes approximately 30,000 deaths annually. Brucellosis costs India at least Rs. 350 million annually. Zoonoses cause considerable loss of livestock, dairy products and protein food by affecting animal population. Thus human health inextricably linked to animal health and production.

Chapter 2

Classification of Zoonoses

There are several zoonotic diseases which have certain reservoirs as well as pattern of their transmission. Zoonoses have been classified on different bases-

Based on Etiological Agent

On the basis of type of etiological agent of a zoonotic disease, the zoonoses are classified into the following types-

1. **Bacterial:** *e.g.,* anthrax, brucellosis, leptospirosis *etc.*
2. **Viral:** *e.g.,* rabies, Japanese encephalitis, Kyasanur forest disease *etc.*
3. **Rickettsial:** *e.g.,* Q fever, scrub typhus, murine typhus *etc.*
4. **Mycotic:** *e.g.,* histoplasmosis, dermatophytosis, coccidioidomycosis *etc.*
5. **Parasitic:** *e.g.,* taeniasis, hydatid disease, visceral larva migran *etc.*

Based on Maintenance Cycle of Infectious Agent in the Nature (Transmission Cycle)

Infectious agents of zoonotic diseases may require different type of hosts and their number and the way of transmission. On this basis of maintenance cycle of infectious agent in the nature, zoonoses are classified into four main types:

1. Direct Zoonoses

In these types of zoonoses there is requirement of single vertebrate species for perpetuation of the infectious agent, *e.g.,* anthrax, brucellosis, leptospirosis, literiosis, rabies, ringworm *etc.*

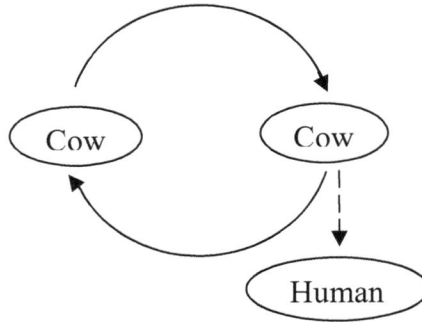

Figure 2.1: Direct Zoonoses (*e.g.*, Brucellosis).

Note: Solid-line represents obligatory pathway and broken-line represents alternative pathway.

2. Cyclozoonoses

These require more than one vertebrate species but no invertebrate host is required for the completion of developmental cycle of the infectious agent. Cyclozoonoses are further classified on the basis of involvement of human beings in the development of infectious agent.

Cyclozoonoses Type I (Obligatory Cyclozoonoses)

These types of zoonoses require human for the development of the infectious agent. Here the involvement of human is compulsory, *e.g.*, taeniasis (*Taenia solium* and *Taenia saginata*).

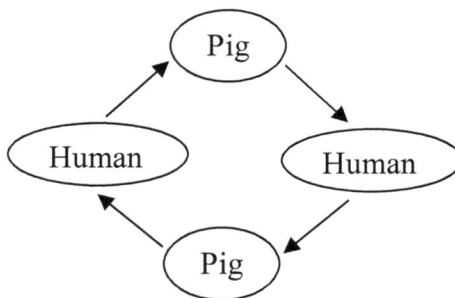

Figure 2.2: Obligatory Cyclozoonoses (*e.g.*, *Taenia solium* infection).

Cyclozoonoses Type II (Non-obligatory Cyclozoonoses)

The involvement of human being is not essential. However, man sometimes involves, *e.g.*, hydatid disease.

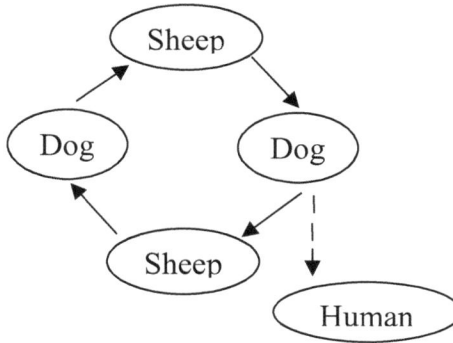

Figure 2.3: Non-obligatory Cyclozoonoses (*e.g.*, echinococcosis).

3. Metazoonoses

Metazoonoses require both the vertebrates and invertebrates and transmitted biologically by invertebrate vectors in which the infectious agent multiplises (plague), develop (Malayan filaria) or both (babesiosis). On the basis of host required, metazoonoses are classified into four types:

Metazoonoses Type I

These require one vertebrate and one invertebrate species, *e.g.*, yellow fever, Japanese encephalitis *etc.*

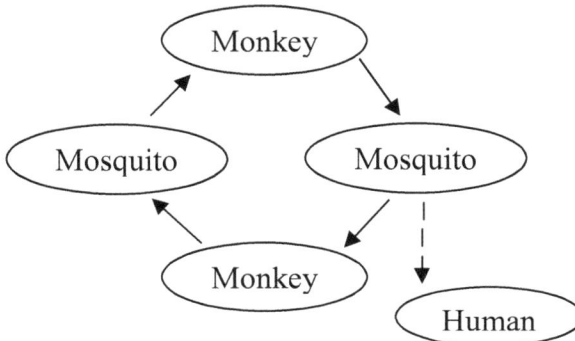

Figure 2.4: Metazoonoses Type I (*e.g.*, yellow fever).

Metazoonoses Type II

These require one vertebrate and two invertebrate species, *e.g.*, paragonimiasis.

Metazoonoses Type III

These require two vertebrate and one invertebrate species, *e.g.*, clonorchiasis.

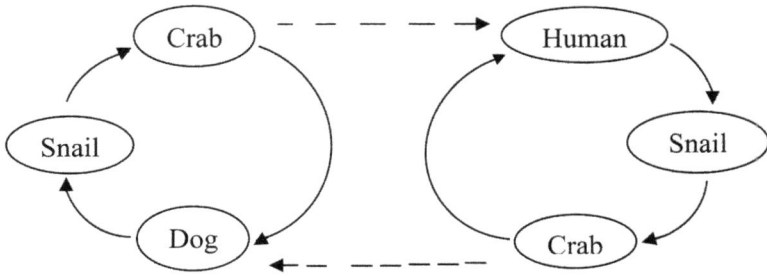

Figure 2.5: Metazoonoses Type II (*e.g.*, paragonimiasis).

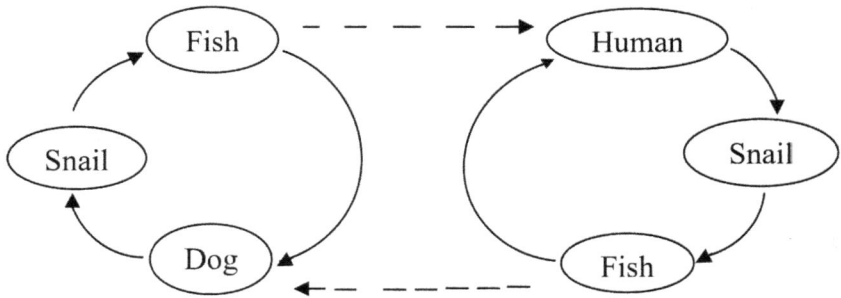

Figure 2.6: Metazoonoses Type III (*e.g.*, clonorchiasis).

Metazoonoses Type IV

These represent transovarian transmission of the infectious agent. The infectious agent is passed transovarially to the vectors progeny, *e.g.*, tick-borne encephalitis (European tick-borne encephalitis, Russian spring summer encephalitis *etc.*).

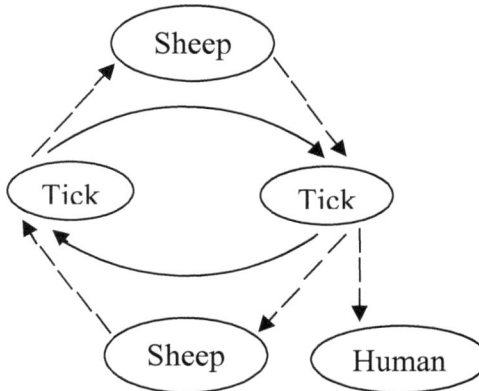

Figure 2.7: Metazoonoses Type IV (*e.g.*, tick-borne encephalitis).

4. Saprozoonoses

The infectious agent requires an inanimate reservoir (food, water, soil, plants, clothing *etc.*) for completion of life cycle in addition to vertebrate or invertebrate. An infectious agent may multiply or develop in inanimate reservoir. Saprozoonoses are classified into three types:

Saprozoonoses Type I (Sapro-amphixenoses)

These diseases are equally shared by animals or humans that are transmitted from animals to humans and vice versa. The agent may undergo essential development in non-animal site, *e.g.*, histoplasmosis (*Histoplasma capsulatum*) in which the agent develops in the soil enriched by bird or bat manure.

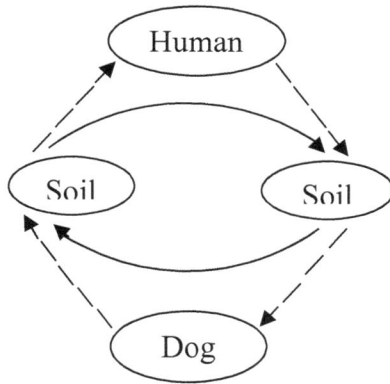

Figure 2.8: Saprozoonoses Type I (*e.g.*, histoplasmosis).

Saprozoonoses Type II (Sapro-anthropozoonoses)

These diseases are transmitted from animals to humans via inanimate reservoirs. The agent may propagate or multiply in non-animal site, *e.g.*, cutaneous larva migran (*Ancylostoma braziliense* infection) in which the agent develops in the soil.

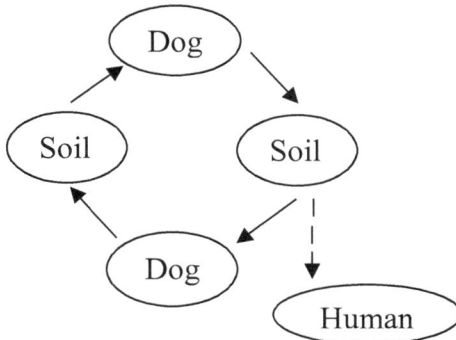

Figure 2.9: Saprozoonoses Type II (*e.g.*, *Ancylostoma braziliense* infection).

Saprozoonoses Type III (Sapro-meta-anthropozoonoses)

The agent requires vertebrate, invertebrate and inanimate reservoir for completion of transmission cycle, *e.g.*, fascioliasis.

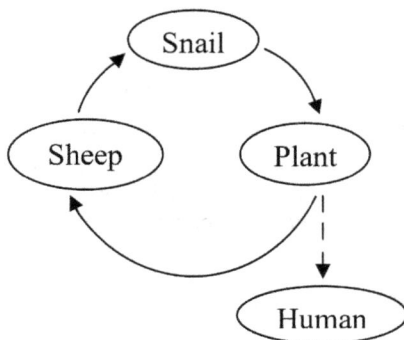

Figure 2.10: Saprozoonoses Type III (*e.g.*, fascioliasis).

Based on Direction of Transmission (Reservoir Host)

Zoonotic pathogens have their certain reservoirs which may be animals or humans. On the basis of direction of transmission (reservoir host), zoonoses are classified into three types:

1. Anthropozoonoses

The type of zoonoses in which the agent is transmitted from animal to human, *e.g.*, anthrax, brucellosis, plague, rabies, hydatid disease, trichinellosis, toxoplasmosis, visceral larva migrant (*Toxocara canis, T. cati* and *Angiostrongylus cantonensis*) etc.

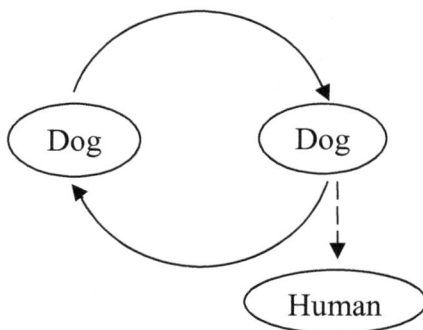

Figure 2.11: Anthropozoonoses (*e.g.*, rabies).

2. Zooanthroponoses

The type of zoonoses in which the agent is transmitted from man to animal, *e.g.*, diphtheria, human tuberculosis, amebiasis, poliomyelitis, infectious hepatitis, legionnaire's disease *etc.*

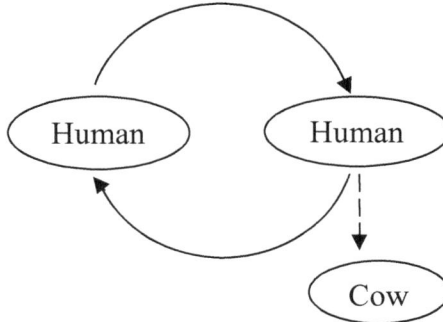

Figure 2.12: Zooanthroponoses (*e.g.*, human tuberculosis).

3. Amphixenoses

The type of zoonoses in which the agent may be transmitted from animal to human or vice versa, *e.g.*, staphylococcosis, streptococcosis, salmonellosis, *E. coli* infection, *Schistosoma japonicum* infection, Chagas' disease *etc.*

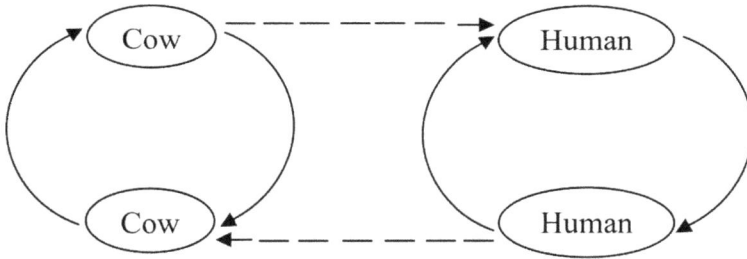

Figure 2.13: Amphixenoses (*e.g.*, staphylococcosis).

Based on Infection Sharing

Depending on the type of animal involved in the transmission of disease, the zoonoses are classified into three types:

1. **Domestic animal-man shared zoonoses:** *e.g.*, anthrax, brucellosis, tuberculosis *etc.*
2. **Wild animal-man shared zoonoses:** *e.g.*, rabies, Kyasanur forest disease *etc.*
3. **Domicilated animal-man shared zoonoses:** *e.g.*, plague.

Chapter 3

Role of Domesticated Pets, Wild and Cold-Blooded Animals in Transmission of Zoonotic Diseases

Zoonotic pathogens have their different types of reservoirs such as domestic, wild and cold-blooded animals. They are potential sources of zoonotic pathogens and play important role in disease transmission. The domestic, pet and wild animal reservoirs of some important zoonotic diseases are as follows:

Cattle

Brucellosis, campylobacteriosis, leptospirosis, tuberculosis, Enterohemorrhagic *E. coli,* cryptosporidiosis *etc.*

Sheep and Goat

Leptospirosis, Russian spring summer encephalitis, cryptosporidiosis *etc.*

Dog

Ehrlichiosis, leptospirosis, rabies, *Brugia* filariasis, cutaneous larva migrans *etc.*

Cat

Cat scratch disease, toxoplasmosis, *Brugia* filariasis *etc.*

Pig

Erysipeloid, leptospirosis, salmonellosis, yersiniosis, Japanese encephalitis, hepatitis E virus infection *etc.*

Horse

Glanders, leptospirosis, Venezuelan equine encephalomyelitis *etc.*

Rodents

Leptospirosis, murine typhus, plague, rat bite fever, cryptosporidiosis *etc.*

Monkey

Kyasanur forest disease, yellow fever, simian malaria *etc.*

Birds

Campylobacteriosis, Japanese encephalitis, histoplasmosis, cryptosporidiosis *etc.*

Role of Domesticated Pets in Transmission of Zoonotic Diseases

Approximately two-third of domestic animals are reservoirs of widespread zoonoses. The role of domestic animals in epidemiology of many zoonotic diseases particularly the direct zoonoses and cyclozoonoses are well understood.

Role of Wild Animals in Transmission of Zoonotic Diseases

Wild animals and birds are the known hosts to a large number of different microorganisms. Many infections of man and domestic animals exist silently in wild species as infections are not apparent. Animals such as monkey, rats, mice *etc.* play important role in epidemiology of many zoonotic diseases. Only a few other wild animal species such as rabid bats, foxes, wolves, skunks *etc.* used as food very often transmit infections directly to human. The monkey acts as reservoir of Kyasanur forest disease, yellow fever, Venezuelan equine encephalomyelitis, California encephalitis *etc.* Hoogstraal (1961) observed the possible role migratory birds in dissemination of ectoparasite vectors of tick typhus (Boutonneuse fever), Crimean hemorrhagic fever, Q fever *etc.* between Africa and Europe or Asia. He also speculated that introduction of Kyasanur forest disease in India was due to migratory birds and tick vectors (*Haemaphysalis spinigera*) which they share various pet animals. Available evidence also suggests that the migratory birds have involved in the spread of West Nile fever and Western and Eastern equine encephalomyelitis viruses.

Role of Cold-Blooded Animals in Transmission of Zoonotic Diseases

Large number of saprophytes or facultative saprophytes has been isolated from the tissues of cold-blooded vertebrates. Fish has been involved in the transmission of many zoonotic pathogens. *Exysipelothrix insidosa* which causes occupational infection (erysipeloid) among meat and fish handlers has been isolated from superficial slime of marine fish. Contaminated fish meal has been incriminated as the source of infection in several outbreaks of erysipeloid in swine. *Clostridium botulinum* type E and *Clostridium chauvoei* has also been isolated from fish. *Listeria monocytogenes* has been isolated from fish and fish products. *Salmonella* spp. has been isolated from fish, frog, snake, lizard and tortoise. *Brucella abortus* has been isolated from toad, snake, lizard and tortoise. *Leptospira* spp. has been isolated from lizard and snake. Rhabdovirus has been isolated from fish and reptiles. *Histoplasma* spp. has been isolated from frog and lizard. *Coxiella burnetii* has been isolated from reptiles. Therefore, they act as important source of zoonotic pathogens.

Chapter 4

Transmission of Zoonotic Diseases

Transmission of zoonotic diseases occurs through various modes. It depends on type of organism and source of infection.

Source of Infection

Any host organism, object or substance from which an infectious agent passes to a susceptible individual is the source of infection. Source of infection may be of following types:

1. **Vehicle:** It is any non-living thing or substance (food, water, milk, dust, serum, pus *etc.*) by which or upon which an infectious agent passes from an infected individual to a susceptible one.

2. **Vector:** It is an invertebrate animal responsible for transmission of infectious agent from an infected individual or its excreta to a susceptible individual (or to some immediate source of infection such as water or food).

3. **Reservoir:** Domestic or wild animals, birds and humans that may or may not show clinical disease but help to maintain the infection for a long period are known as reservoirs.

Classification of Modes of Transmission of Infection

Based on Pattern of Transmission of Infectious Agent in the Nature

Horton-Smith (1957) classified the modes of transmission as follows:

1. **Vertical transmission:** It is the transmission of an infectious agent from an individual to its offspring such as congenital or transovarian transmission, *e.g.*, babesiosis, Russian spring summer encephalitis, Colorado tick fever, Kyasanur forest disease *etc.*

2. **Horizontal transmission:** It is the transmission of an infectious agent from an infected individual to a susceptible contemporary, *e.g.,* tuberculosis, influenza, mange, ringworm *etc.*

3. **Zigzag transmission:** It is an alternate passes of infectious agent from one type of animal (vertebrate) to another type (invertebrate), *e.g.,* Japanese encephalitis, yellow fever *etc.*

Based on Source of Infection

1. **Contact transmission:** It means the transmission of an infectious agent from an infected individual or its secretion to a susceptible individual as follows:

   a. **Actual direct physical contact:** *e.g.,* dourine, syphilis, herpes simplex, some dermatophyte infections *etc.*

   b. **Immediate or short-term indirect contact:** It occurs through contaminated objects.

2. **Air-borne transmission:** *e.g.,* pulmonary tuberculosis, canine distemper, measles *etc.*

3. **Vehicle transmission:** Vehicle transmission may be of following types:

   a. **Mechanical vehicle transmission:** It involves the survival of the agent without its development or propagation, *e.g.,* amebiasis, serum hepatitis *etc.*

   b. **Propagative vehicle transmission:** It involves the propagation of the agent in or on the vehicle, *e.g.,* milk-borne streptococcosis, histoplasmosis *etc.*

   c. **Developmental vehicle transmission:** It involves the essential development of the agent in or on the vehicle, *e.g.,* hookworm disease.

   d. **Cyclopropagative vehicle transmission:** It involves both propagation and development of the infectious agent in or on the vehicle, *e.g.,* strongyloidiasis.

4. **Vector transmission:** Vector transmission may be of following types:

   a. **Mechanical vector transmission:** It involves the survival of the agent in or on the vector without its multiplication or essential development, *e.g.,* typhoid fever, myxomatosis, mechanically transmitted trypanosomes *etc.*

   b. **Propagative vector transmission:** It involves multiplication of the agent in or on the vector, *e.g.,* plague, yellow fever *etc.*

   c. **Developmental vector transmission:** It involves the essential development of an agent in or on the vector, *e.g.,* filariasis.

   d. **Cyclopropagative vector transmission:** It involves both multiplication and development of an infectious agent in or on a vector, *e.g.,* malaria.

   e. **Trans-ovarian vector transmission:** This is a special case of mechanism of propagative vector transmission in which the infectious agent is passed transovarially to the vector's progeny, *e.g.,* relapsing fever in ticks, tick-borne rickettsiosis, piroplasmosis *etc.*

f. **Trans-stadial vector transmission:** It involves the passage of an agent from one stage of development of a vector to another, *e.g.,* rickettsiosis, heart water disease *etc.*

Chapter 5

Zoonotic Pathogens as Agent of Bioterrorism

Bioterrorism

It is a criminal activity against civilian using microorganisms or their toxins derived from them. It is the threat of intentional introduction of a microorganism that can affect animal and human health.

Biologic Warfare

It is defined as the use of pathogenic microorganisms to harm or kill animal and human population and food. Pathogenic microorganisms are being used since ancient time. Many countries have been involved in the activity of bioterrorism.

Biologic Warfare Agents

It includes both the living microorganisms and their toxins. Biologic warfare agents may be bacteria, viruses and fungi.

1. **Bacteria:** *e.g., Bacillus anthracis, Brucella suis, Salmonella* Typhimurium, *Shigella* spp., *Vibrio cholerae, Francisella tularensis, Yersinia pestis, Coxiella burnetii etc.*

2. **Viruses:** *e.g.,* Small pox virus, *Ebola* hemorrhagic fever virus, yellow fever virus *etc.*

3. **Microbial toxins:** *e.g.,* botulinum toxin, staphylococcal enterotoxin B *etc.*

4. **Anti-crop agent:** These include rice blast, rye stem rust, wheat stem rust *etc.*

Criteria of Selection of Agents of Bioterrorism

The infectious agents selected as agent of bioterrorism have the properties such as lethal, highly communicable, persist in the environment, low infective dose, easy to produce and aerosol dissemination in the target areas.

Mode of Delivery of Agents of Bioterrorism

1. **Aerosols:** It is easy for terrorists to release the agent through aerosols. It is also easy to release the infectious agent to large number of population. A bio-weapon aerosol is made invisible and odorless and therefore cannot be detected easily when released.

2. **Water contamination:** This is an ideal target for terrorists to contaminate the water supplies in cities. 0.07-0.1 microgram of botulinum toxin is enough to kill human being.

3. **Food contamination:** The most successful bioterrorist events recognized to date is through food as a vehicle for transmission of infectious agents.

4. **Postal way:** The anthrax attack on the east coast of the US has demonstrated that mailed packages, such as envelope were used as a method of dissemination.

Emergency Measures Against Bioterrorism

Following emergency measures can be adopted against bioterrorism:

1. Establishment of decontamination of affected areas.
2. Containment of spill/release of organisms.
3. Pre-entry examination and determination of appropriate protective clothing and equipment.
4. Keeping of record after action.
5. Complete analysis of action/recommendations to action plan.

Bio-monitoring Against Bioterrorism

Rapid detection and identification of the causative agent are the most crucial step in the management of the events. The conventional microbiological methods like culture, serology and molecular identification are suitable for detection and identification of the agent in the laboratory. The online detection of microorganisms is possible through the identification of various bioreceptor molecules as the signal generating systems *viz.*, firefly lucerferase and green fluorescent protein. Detection of microorganism by biosensor is also useful technique. Biosensor is a type of probe in which the biological components interact with an analyte which is detected into by electric component and translated a measurable electronic signal. Some other methods *viz.*, bioprobe, LASER and electronic masses with incorporated alarms have also been used for detection and identification of agents of bioterrorism.

Diagnosis and Treatment Against Agent of Bioterrorism

The agent of bioterrorism can be detected and identified from the contaminated places/patients according to the clinical symptoms, pattern of disease occurrence and laboratory findings. Contaminated laces/areas must be decontaminated and to make free from agent of bioterrorism in order to remove/destroy the sources of infection. The treatment of patients should be started immediately following the identification of agent.

Chapter 6

Method of Prevention, Control and Eradication of Zoonotic Diseases

Prevention

Prevention implies all measures taken to exclude a disease from unaffected (healthy) population. The first line of defense against disease is prevention. Prevention of a disease in a population can be done by adopting following measures-

1. Quarantine

Quarantine is a restraint placed upon the movement of humans, animals, plants or goods which are suspected being of carriers or vehicle of infection or of having been exposed to infection. In general sense quarantine refers to detention of animals suspected of disease at port or land borders. Quarantine is the oldest of tools of preventive medicine. The aim of quarantine is to prevent the healthy native animals from coming in contact with imported infected animals or biomaterials. The first international livestock quarantine imposed by the United States was put into effect in 1890. Under this provision all imported cattle were held at port-of-entry quarantine stations for 90 days from their date of shipment. All ship and swine were held for 15 days from their date of arrival. The quarantine period resembles in some respects the 10 days observation period as granted to a suspected case of rabies in dog. In case of rabies, however, it is not called quarantine as the incubation period can be much longer. Quarantine period was originally fixed as 40 days but now days it is left to the discretion of the appropriate authorities, who in most of the cases are veterinary officer. As per model rules the period of quarantine can be extended to 90 days. Use of quarantine in public health use has been chiefly at the international level. As for as man is concerned, International Sanitary Regulations are directed against six so-called quarantinable diseases as yellow fever, small pox, cholera, plague, louse-borne typhus and louse-borne relapsing fever.

2. Mass Immunization

The second preventive tool is the mass immunization of population. Mass immunization as a preventive technique has certain definite advantages over alternative procedures as well as some disadvantages.

Advantages of Mass Immunization

i. The ability of the restraint individual to move about freely, always carrying his protection with him (as not the case with environmental measures).

ii. His ability to dispense with frequently repeated doses which are necessary in chemoprophylacsis (except in some instances for occasional "boosters").

Disadvantages of Mass Immunization

i. Lack of lifetime protection in man in most instances.

ii. The imperfect protection afford by vaccines.

iii. In addition, there are certain actual or potential risks associated with use of some vaccines include sensitization as for example to kidney tissue in the case of monkey kidney tissue culture vaccine, inoculation of tumor inducing chromosomal material, unwanted reaction to foreign nervous tissue (including perhaps in some cases multiple sclerosis like disease) and the possibility of spread of viruses of foot and mouth disease, encephalomyocarditis and lymphocytic choriomeningitis.

3. Environmental Measures

Third group of preventive technique is measures for environmental control. In addition to measures to ensure safe water supplies, environmental measures include air sanitation, food protection and control, pest and vector control, improvement of housing; sanitation of swimming pools, bathing beaches *etc.* (in the case of human), excrement treatment and disposal and garbage disposal.

4. Chemoprophylaxis

It includes the use of chemicals in the prevention of illness. Progress has been realized in preventing rheumatic fever through prophylactic administration of penicillin on a regular schedule to persons who have had streptococcal infection. It is also useful as repellents against both ectoparasites and insect vectors.

5. Early Detection

Early detection is a method specially suited to chronic infectious diseases and diseases of non infectious nature. The purpose of early detection is to diagnose an early preclinical stage or to determine the state of immunity of the population. The use of tuberculin test first in domestic animal populations then in human populations is a classic example of early detection of tuberculosis. There are some other tests which can be used for the early detection of particular disease, *e.g.,* brucellin test, mallein test, histoplasmin test, sporotrichin test, Casoni test and Widal test for brucellosis, glanders, histoplasmosis, sporotrichosis, hydatid disease and *Salmonella* infection, respectively.

6. Health Education

This is the most effective preventive device. It includes see the posters, notices, handbills so crowded with small type. Education to be most effective as a preventive tool, must work towards engendering an assumption of initiative disease prevention.

Control

Disease control is a broad term generally used to describe all measures taken to combat illness among members of a population and to eliminate cause of illness which may exist in the environment. In the instance of infectious disease, disease control implies a reduction in the number of cases or in the opportunities for transmission to a level where the infection no longer exist as a major public health or economic problem. There are several methods by which the disease can be controlled.

1. Test and Slaughter

In this method the affected animals are detected through mass surveys and removed from the population by subjecting them to premature slaughter. Stray dog elimination has been an effective measure against rabies and hydatid disease. Test and slaughter is not initially practical where the prevalence of infection is very high or in instances where the availability of replacement stock presents a real problem. The principal disadvantages of this approach are its high initial cost in operation and in compensation and the fact that it is often difficult approach to "sell" to farmers.

2. Mass Treatment

Mass treatment of the affected population may be carried out either in emergency situation or when disease prevalence is very high and the drug used is very safe. To be effective, mass treatment must be applied either prophylactically or curatively on a thorough population basis. Early uses of this tactic were routine addition of coccidiostats to poultry drinking water, routine incorporation of anthelmintic into ruminant salt licks or feeds, use of some nutritional supplement mixes and some therapeutic incorporation of antibiotics in feeds. In the case mass antibiotic use, development and spread of drug resistant microorganism is possible. The difficulties of mass treatment approaches are that they require the cooperation of an affected human population or of a population of livestock owners.

3. Vector Control

The general principles of vector control are as follows:

Environmental Control

It includes elimination of breeding places, filling and drainage operation, careful planned water management, provision of piped water supply, proper disposal of refuse and other wastes and cleaning in and around house. Intensive health educations of the public as well as political supports are essential prerequisites.

Vegetation clearance has proven as effective control against chrysops, loa loa and tse tse fly vectors of the trypanosomes. Vegetation clearance along with water courses has been locally effective in both anopheline mosquito and snail control.

Chemical Control

A wide range of insecticides belonging to the organophosphorus (abate, chlorpyrefos, dichlorvos, malathion, parathion *etc.*), organochlorines (aldrin, dialdrin, BHC, DDT, lindane, methoxychlor *etc.*) and carbamates (cabaryl, dimetilan, pyrolan, propoxur *etc.*) group of compounds are available for vector control. Insecticide application has been a particularly difficult matter in instance where vectors breed in swiftly flowing water, as do simulidae or in densely vegetated areas as does chrysops in West Africa. To avoid undue environmental pollution, it is now considered essential to replace the highly persistent compounds with readily "biodegradable" (*e.g.*, dursban, abate and methoxychlor) and less toxic to human and animals. Demerits of using insecticides for vector control include development of resistance in many arthropod vectors and environmental pollution.

Biological Control

Biological control means control of vectors, reservoir host or a parasite by the introduction of natural enemies into a habitat. To minimize the environmental pollution with toxic chemicals, great emphasis is now placed on biological control. There are various natural enemies effective against certain vectors. *Bacillus sphaericus, Bacillus thuringiensis* var *israelensis* and pathogenic microsporidian such as *Thelokomia opecita* have been used for control of mosquitoes. Larvivorous fish like *Gambusia affinis, Lebister reticulatus (Barbados millions)*, Barbel fish and Giant gourami fish are important natural enemies against vectors like mosquitoes and cyclops. Growing of the water fern *Azolla microphylla* in rice field has been found as a biological agent against mosquitoes breeding in rice fields. Moreover, some bacteria and parasites have been successful against snail vectors which resulted into reduced schistosomiasis. A variety of biological agents like bacteria, viruses, fungi, nematodes and protozoa are under study for control of insects. But the fear exists that the introduction of biological agents for the control of arthropods may pose a direct hazard to the health of human himself. Therefore, the natural enemies used for control of vector must possess some important characteristics such as harmless to the animals and humans, economically feasible, easy and safe to handle and to maintain, simple and convenient for their applications, do not cause environmental nuisance and effective enough to be used as dependable means of vector control.

Genetic Control

The WHO/ICMR research unit at New Delhi has contributed massively to the techniques of genetic control of mosquitoes. The techniques such as sterile male techniques, cytoplasmic incompatibility, and chromosomal translocation have been found to be effective in small field trials.

Newer Methods

New and innovative methods such as chemosterilant and sexattractants or pheromones are being sought for pest control.

Integrated Approach

Since no single method of control is likely to provide a solution in all situations, the present trend is to adopt an " Integrated approach" for vector control combining two or more methods with a view to obtain maximum results with the minimum effort and to avoid the excessive use of any one method.

4. Reservoir Control

This approach has met with great success in measures directed against rats, stray dogs and other noxious reservoir hosts of infection such as leptospirosis, plague, typhus and rabies. Stray dog elimination has been an effective measure against hydatid disease and rabies. Trapping or poisoning of foxes as well as other wild host is also helpful in controlling rabies. Poison baiting and trapping have been among the most commonly employed technique against reservoirs. Poison baiting is potentially dangerous because of risk to children, pets and other animals. Use of rodenticide is helpful in controlling of zoonotic infections like leptospirosis and plague. Antirodent measures are applicable only when a population of expendable wild animals acts as the reservoir for an infection of domestic animals or of human. Rodents can be controlled by adopting following measures:

Sanitation Measures

It includes proper storage, collection and disposal of garbage; proper storage of food stuffs; construction of rat proof buildings, godowns and warehouses; and elimination of rat burrows by blocking them with cement and concrete.

Trapping

The traps are usually baited with indigenous foods of the locality. The captured rats must be destroyed which may be done by drowning them in water.

Fumigation

Fumigation is effective against both rat and rat fleas. The fumigants which can be used are calcium cyanide (often called cynogas or cymag), carbon disulphide, methyl bromide, sulpher dioxide *etc.* Cynogas has been extensively used in India for the fumigation of rat burrows. This chemical is prepared in powder form and is pumped into rat burrows by a special foot pump called the "cynogas pump".

Chemosterilants

Chemosterilants cause temporary or permanent sterility in either sex or both sexes. Rodent chemosterilants are still in the experimental stage.

Rodenticides

An expert committee of WHO grouped the acute rodenticides into three groups. Those requiring ordinary care include red squill, norbromide and zinc phosphide. Those requiring maximal precautions include strychnine, fluorocetate and fluoroacetamide. Those are too dangerous for use includes alpha naphthyl thiourea (ANTU), arsenic trioxide, gophacide, phosphorous and thallium sulphate. The commonly used rodenticide in India are zinc phosphide and barium carbonate.

The use of rubber gloves are recommended in handling zinc phosphide as it is highly poisonous. Because of its good safety record, low cost and reasonably high effectiveness, zinc phosphide is recommended on large scale use against rats. Rodenticides are of two main types.

Single Dose (Acute)

It is lethal to the rat after single feeding.

Multiple Doses (Cumulative)

It requires repeated feeding over a period of 3 or more days. It includes warfarin, pindone, diphacinone, and coumafuryl.

5. Miscellaneous Measures

There are a number of other specialized approaches by which the disease can be controlled by using heat, cold irradiation and other physical and chemical sterilization procedures.

Eradication

The third strategy of population medicine is that of disease eradication. The literal meaning of eradication is "tear out by roots". It means complete removal of a disease from an area or country. Eradication implies action and completeness; it is not eradication if a single individual of the parasite species remains. It is quite obvious, however, that for some infections no possibility of eradication is conceivable because of nature of the reservoir as in yellow fever or the ubiquitousness of the parasite as in certain of the salmonellosis and staphylococoosis. Today, small pox is the only disease that has been eradicated during recent years; three diseases have been seriously advanced as candidates for global eradication within the foreseeable future. Such diseases are polio, measles, and dracunculiasis.

Species Eradication as the Principle

The extinction of parasite species which pray on man and domestic animals is an exciting prospect. It has been occurred many times through purely natural forces. In considering species eradication, however, we are implying a deliberate effort to derange drastically an organism's ecology in order to bring about its disappearance. There are two successful technical approaches to the regional eradication of infections, one is the test and slaughter and other is the vector control. Mass immunization has also been used in the regional eradication of a few infections such as small pox in human.

1. Test and Slaughter (Reservoir Control)

This method of eradication requires a reliable diagnostic test which can be applied on mass basis. With the diagnosis made, the next step in test and slaughter are the establishment of a local quarantine of infected premises and the initiation of ecological studies to ascertain the extent of infection. The next step in stamping out the disease is immediate slaughter of all infected and exposed animals and their disposal by deep burial in quicklime or by burning. The premises and all possible contaminated objects and materials are then subjected to thorough cleaning and disinfection. In developed countries many of the major zoonoses such as tuberculosis, brucellosis, rabies and hydatid disease have been successfully eradicated by adopting test and slaughter policy.

2. Vector Eradication

The origin of the concept of vector eradication can be pinpointed to a paper written by Cooper Curtice of the US Bureau of Animal Industry (BAI) in 1891, the first of a number of papers he wrote on the subject of tick eradication and he is known as "Father of tick eradication".

3. Mass Immunization

This technique has also been used in the regional eradication of a few infections such as small pox in human.

Section - II

Bacterial Zoonoses

Over 200 zoonotic diseases have been described. Several bacterial zoonotic diseases are known since centuries. Bacterial zoonoses are one of the important zoonoses caused by pathogens. There are various emerging and re-emerging zoonotic diseases among which bacterial zoonoses contribute significantly. Bacterial zoonoses cause morbidity and mortality in the animals and humans, and economic losses in the form of treatment, hospitalization, loss of wages and reduced production in the animals. Almost 100years ago, prior to application of hygiene rules and discovery of neither vaccines nor antibiotics, some bacterial zoonotic diseases such as bovine tuberculosis, bubonic plague, and glanders caused millions of human deaths. Moreover, millions of people get sick every year because of foodborne zoonoses such as salmonellosis and campylobacteriosis which cause fever, diarrhea, abdominal pain, malaise and nausea. Other bacterial zoonoses are anthrax, brucellosis, infection by verotoxigenic *Escherichia coli*, leptospirosis, plague, Q fever, shigellosis and tularemia. Most of the bacterial zoonoses are transmitted directly, while some bacterial zoonoses are transmitted by vectors (plague, trench fever, carrion's disease, Lyme borreliosis, relapsing fever, ehrlichiosis, rickettsial diseases *etc.*). The development of antimicrobial resistance due to over/misuse of antibiotics is also a globally increasing public health problem. These diseases have a negative impact on travel, commerce, and economies worldwide. In most industrialized countries, antibiotic resistant zoonotic bacterial diseases are of particular importance for at-risk groups such as young, old, pregnant, and immune-compromised individuals. The closer contact with companion animals and rapid socioeconomic changes in food production system has increased the number of animal-borne bacterial zoonoses. The spread and importance of some bacterial zoonoses are currently increasing worldwide. Some important bacterial zoonoses are discussed in next chapters.

Chapter 7

Anthrax

Synonyms

Wool sorter's disease, Splenic fever, Charbon, Milzbrand, Malignant pustule

Anthrax is an acute septicemic and contagious disease of mammals especially the herbivores, although few, if any, warm blooded species are entirely immune to it. It is characterized by fever, splenomegaly and bleeding from natural orifices. Anthrax is an occupational anthropozoonosis affecting agricultural, animal husbandry and industrial workers.

Etiology

Anthrax is a bacterial disease caused by *Bacillus anthracis*, a Gram-positive, rod-shaped, capsulated, aerobic and spore-forming bacterium. Capsules are formed in the body of animals. Sporulation requires the presence of free oxygen; within the anaerobic environment of the infected host the organism is in the vegetative form. The spores of the organism are very much resistant to cold, hot, chemicals and drying. The spores may remain viable in the soil for long period and for 10 years in the infected tissues and cultures. The spores remain resistant to 100°C for 5 minutes. A 5 per cent sodium hydroxide can effectively destroy the spore contaminated objects. High humidity, alkaline soil and high amount of organic matter provide ideal conditions for the propagation and survival of *B. anthracis*. Such sites are called "incubator areas" of anthrax. The organism is discharged in the blood and excretions of the animal on death. In intact carcases, vegetative cells get destroyed on account of putrefactive changes.

The capsule and the toxin complex are the two known virulence factors of *B. anthracis*. The poly-D-glutamic acid capsule is presumed to act by protecting the bacterium from phagocytosis. The toxin complex, which consists of three synergistically acting proteins like Protective Antigen (PA), Lethal Factor (LF) and

Edema Factor (EF) are produced during the log phase of growth of *B. anthracis*. LF in combination with PA (lethal toxin) and EF in combination with PA are now regarded as responsible for the characteristics signs and symptoms of anthrax.

Epidemiology

The disease occurs worldwide. The Office International des Epizooties (OIE) reported that the disease is still enzootic in most countries of Africa and Asia, a number of European countries and countries/areas of the American continent and certain areas of Australia; it still occurs sporadically in many other countries. Anthrax is enzootic in southern India but is less frequent to absent in the northern Indian states where the soil is more acid, while in Nepal it is endemic. Some Southeast Asian countries (Myanmar, Vietnam, Cambodia, and Western China) are severely affected, while Thailand used to be free. Malaysia and Taiwan are free, and the disease is limited to single regions of the Philippines and Indonesia. In India the disease is enzootic in nature and occurs in all the districts of different states.

Anthrax is primarily a disease of herbivores. However, the disease has also been reported in dogs scavenging anthrax carcasses and in carnivorous animals in zoological gardens and wildlife sanctuaries or parks, though outbreaks affecting large numbers of carnivorous animals are very rare. There has been a progressive global reduction in livestock cases in response to national programmes. In humans, anthrax generally occurs as sporadic.

Transmission

The major sources of human anthrax infection are direct or indirect contact with infected animals, or occupational exposure to infected or contaminated animal products. Other possible sources are rare and epidemiologically trivial. Except when taken up by the pulmonary route, *B. anthracis* needs a lesion through which to enter the body. Following entry, the spores, which may have commenced germination, are carried to the lymphatics where they multiply. The disease may be transmitted through the following routes-

1. **Contact:** Contact with blood and discharges from infected animals and contaminated soil, sewage, water, animal feeds and animal byproducts.
2. **Ingestion:** Ingestion of raw or undercooked meat or dairy products from infected animals.
3. **Inhalation:** Inhalation of spores, contaminated dust or wool.

Disease in Animals

The incubation period in the susceptible herbivores ranges from about 36 to 72 hours. It occurs in three forms as follows:

1. Per-acute Form

It is characterized by septicemia and sudden death. The hyper-acute systemic phase is characterized usually without easily discernible prior symptoms. The first signs of an anthrax outbreak are one or more sudden deaths in the affected livestock,

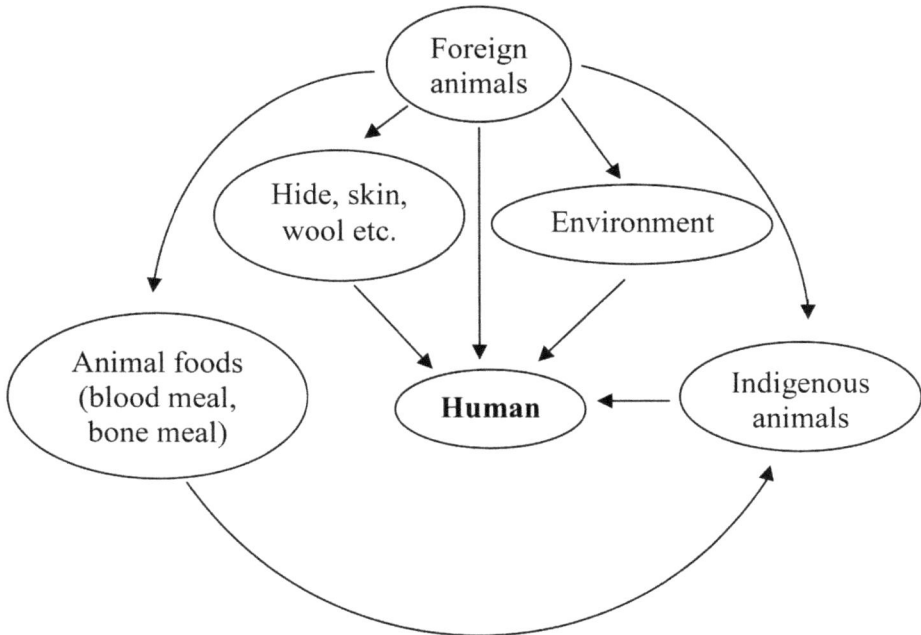

Figure 7.1: Infection Cycle of Anthrax.

although farmers may reflect retrospectively that the animals had shown signs such as having been off their food or having produced less milk than usual. During the systemic phase, the animals become distressed, appear to have difficulty in breathing and cease eating and drinking. Swellings in the sub-mandibular fossa may be apparent; temperatures may remain normal for most of the period or may rise. The animal can remain responsive to treatment well during this period but if treatment fails it lapses into coma followed by death from shock. In highly susceptible species, the period between onset of visible symptoms and death may be just a few hours; the course of these events is more protracted in more resistant species.

2. Acute Form

It is characterized by fever, extreme depression, increased respiration and pulse rates and extreme dyspnea. After death, there is oozing of blood from natural orifices (nose, mouth, ear and anus). Carbuncles are located on the neck, chest, abdomen and limbs in equine. Hemorrhagic spots are found on the skin surface in swine.

3. Sub-Acute Form

It is characterized by edema under the neck, brisket region, thorax, abdomen and flank.

Disease in Humans

The disease in humans is characterized by three major forms that are cutaneous, pulmonary, and gastrointestinal forms, depending upon mode of entry and localization of the infection. All three forms are potentially fatal if untreated. However, some other forms of anthrax are also noticed. The cutaneous form is often self-limiting. Overt gastrointestinal tract and pulmonary cases are more often fatal, largely because they go unrecognized until it is too late for effective treatment. Development of meningitis is a dangerous possibility in all three forms of anthrax. Both meningitis and septicemia may be complication of any of the three forms. Anthrax may occur in other forms such as anthrax meningitis and anthrax sepsis. Anthrax meningitis (meningeal form) is believed to be sequelae of cutaneous anthrax.

1. Cutaneous Anthrax

The incubation period ranges from 9 hours to 2 weeks (mostly 2 to 7 days). It accounts for about 95-98 per cent of the human cases. This form occurs where the infectious organisms enter a cut or abrasion of the skin. It does not penetrate the intact skin. During the incubation period, organisms germinate, multiply and produce toxin which causes development of small papule. As the primary lesion develops into a fluid filled vesicle, extensive edema may involve in the adjacent tissues. Satellite vesicles may appear near the initial lesion and vesicular fluid initially clear becomes dark and bluish black. When the vesicle ruptures there is necrosis being at center eventually develop into the typical black scar. The lesion is relatively painless until pressure is applied. The symptoms are accompanied with malaise and 2°F rise in temperature along with occasional tenderness of lymph nodes due to lymphadenitis and lymphadenopathy. After development of the scar about 7-10 days after onset, about 80 per cent of the cases progress to healing. The edges of the scar loosen and eventually the entire black scab fall off, leaving the lesion includes heals by granulation results in distinct scar. In about 20 per cent cases there is a progressive extension of infection into the regional lymph nodes and blood that rapidly developed into a bacteremia and toxemia.

2. Pulmonary Anthrax

This is the rarest and most severe form of anthrax. This occurs when viable spores are deposited in the alveoli of the lungs, the spores are phagocytized and carried to the mediastinal lymph nodes, where they germinate, multiply and produce toxin. The most common post-mortem lesions seen are necrotic, hemorrhagic and edematous mediastinal lymph nodes. Pulmonary lesions are minimal or if present are due to a secondary pneumonia extending from mediastinal lymph nodes. The clinical manifestations are divided into two stages:

First Stage

This is characterized by sudden onset of symptoms of mild upper respiratory tract infection accompanied by mild fever, malaise, myalgia and non productive cough. These signs closely mimic influenza, viral respiratory disease or mold bronchopneumonia and thus diagnosis are made.

Second Stage

This is characterized by sudden onset of acute dyspnea, diaphoresis and cyanosis. The temperature is usually elevated but may be depressed due to shock. Pulse and respiration rates are accelerated. Death usually occurs within 24 hours of onset of second stage.

3. Gastrointestinal Anthrax

There are two clinical forms of gastrointestinal anthrax which may occur following ingestion of *B. anthracis* in contaminated food or drink.

Oropharyngeal Anthrax

The main clinical features are sore throat, dysphagia, fever, regional lymphadenopathy in the neck and toxaemia. Even with treatment, the mortality is about 50 per cent.

Intestinal Anthrax

Organism penetrates the intestinal mucosa; the organism then multiplies and produces toxins in the sub-mucosal tissues which cause ulceration. Symptoms include nausea, vomiting, fever, abdominal pain, haematemesis, bloody diarrhea and massive ascites. Unless treatment commences early enough, toxaemia and shock develops, followed by death. There is evidence that mild, undiagnosed cases with recovery occur.

4. Anthrax Meningitis

Meningitis due to anthrax is a serious clinical development which may follow any of the other three forms of anthrax. The case fatality rate is almost 100 per cent ; the clinical signs of meningitis with intense inflammation of the meninges, markedly elevated CSF pressure and the appearance of blood in the CSF (the meningitis of anthrax is a haemorrhagic meningitis) are followed rapidly by loss of consciousness and death. Only a few instances of survival as a result of early recognition of the problem and prompt treatment are on record.

5. Anthrax Sepsis

Sepsis develops after the lymphohematogenous spread of *B. anthracis* from a primary lesion (cutaneous, gastrointestinal or pulmonary). Clinical features are high fever, toxaemia and shock, with death following in a short time.

Diagnosis

The definitive diagnosis is obtained by visualisation of the capsulated bacilli in the CSF and/or by culture. Differential diagnosis should take into account of acute meningitis of other bacterial etiology. Anthrax can be diagnosed by the following methods:

1. Case History and Clinical Examination

The history is of major importance in the diagnosis of anthrax. Clinical manifestations in ruminants are sudden death, bleeding from orifices, subcutaneous haemorrhage, without prior symptoms or following a brief period of fever and disorientation should lead to suspicion of anthrax. In equines and some wild herbivores, some transient symptoms are fever, restlessness, dyspnoea and agitation. In pigs, carnivores and primates the manifestations are local edemas and swelling of face and neck or of lymph nodes, particularly mandibular and pharyngeal and/or mesenteric.

2. Cultural Examination

B. anthracis can be cultured from the blood or from a swab of an ear clipping or other appropriate specimen on blood agar or nutrient agar. The hemorrhagic nasal, buccal or anal exudate will also carry large numbers of *B. anthracis* which can be cultured from swabs or from samples contaminated with the exudate. Knisely's PLET (Polymyxin Lysozyme EDTA Thallous acetate) medium is a selective medium for the isolation of *Bacillus anthracis*. This medium inhibits the growth of most strains of *Bacillus cereus, B. subtilis,* other *Bacillus* species, enterobacteria and *Pseudomonas* spp. On PLET agar, colonies of *B. anthracis* are similar to, but a little smaller than, those nutrient agar (grayish, granular discs, uneven surface and wavy margin *i.e.* "medusa-head" appearance). Culture on blood agar or nutrient agar after overnight incubation shows typical "medusa head" colonies with little or no hemolysis on blood agar.

3. Morphological Examination

A blood smear should be obtained with a swab from a small incision in the ear or from an ear clipping (the ear is usually recommended as being accessible, supplied with an extensive capillary network), or by means of a syringe from an appropriately accessible vein (the blood characteristically clots poorly or not at all upon death in anthrax victims and is dark and haemolysed). The smear is dried, fixed and stained with polychrome methylene blue (McFadyean stain). Large numbers of the capsulated bacilli are seen in smears from relatively fresh carcases of most species. It shows long tangled Gram-positive chain of bacilli. The spores are centrally placed.

4. Serological Examination

The thermo-stable antigen precipitin test devised by Ascoli. It is still used in several countries of Europe and the Far East for detecting residual antigens in tissue in which it is no longer possible to demonstrate *B. anthracis* microscopically or by culture. Serological test like precipitation test (Ascoli test: thermo precipitation test) can be preformed. Take infected tissues in slightly acidified saline and heat. Cool and filter. Take 0.5 ml filtrate and mix with same amount of antiserum and observe for development of a precipitate line at the junction of two liquids, which is considered as positive result.

5. Animal Inoculation Test

Inject 0.5 ml clinical material subcutaneously in guinea pig or intraperitonealy in mice. Death occurs within 30-40 hours.

6. PCR Technique

This is useful for detection of *Bacillus anthracis* organism.

Disease Management in Animals

Following the first incident of anthrax in a herd, the remaining animals should be moved immediately from the field or area where the index case died and regularly checked at least three times a day for two weeks for signs of illness (rapid breathing, elevated body temperature) or of sub-mandibular or other oedema. Any animal showing these signs should be separated from the herd and given immediate treatment.

1. Hygienic Measures

Immediately after detection of anthrax during clinical examination of herds or during slaughter or at the time of post-mortem examination, emergent measures are taken to avoid environmental contamination. The carcase and materials in-contact with it are promptly disposed off.

2. Vaccination

In endemic areas, or if there is concern that the outbreak may spread, the herd should be vaccinated. Vaccination can be done with anthrax spore vaccine 1 ml subcutaneously in cattle, sheep, goat, and horse for all age group. Booster dose is given after 6 months.

3. Treatment

Affected animals are treated with antibiotics like penicillin, tetracycline and chloramphenicol. Anti-anthrax serum may be given 100-200 ml intravenously with a course of penicillin.

4. Miscellaneous Measures

The fodder from contaminated pasture should be destroyed and not to be given to the animals. Hide, wools, bone meals *etc.* should be sterilized by gamma irradiation and get certified from competent authority as free from anthrax spores. It is legally required that no suspected carcase should be opened or incised. The natural orifices should be plugged with cotton soaked with disinfectant. Contaminated material including the animal itself should be collected and carried away in a closed vehicle to the site of disposal. The farm buildings, slaughterhouses, exposed hospital equipment and clothing can be disinfected with 5-10 per cent formalin or 5 per cent phenol or bleaching powder. Anthrax exposed materials are best disposed off by burning or incineration. Pastures, soil and incubation areas are treated with 30 per cent peracetic acid which kills the spores. Wools and other fibres can be treated with sodium carbonate.

Disease Management in Humans

1. Vaccination

Alum precipitated cell-free filtrate of Sterne strain or aluminum hydroxide adsorb cell-free filtrate produced from non-capsulating and non-proteolytic culture is used for vaccination purpose.

2. Treatment

Prompt and timely antibiotic therapy usually results in dramatic recovery of the individual or animal infected with anthrax. The drug of choice is penicillin. Tetracycline and chloramphenicol can also be used as alternative drugs.

3. Hygienic Measures

Person handling anthrax infected animals should adopt adequate sanitary measures for their own safety.

4. Other Measure

Prevent the movement of livestock from affected premises during an outbreak. Control dust in industries handling wool or hides. Wash and disinfect (with10 per cent formalin) the wool and hair from endemic areas.

Carcase Disposal

The sporulation of *B. anthracis* requires oxygen and therefore does not occur inside a closed carcase. Regulations in most countries forbid post-mortem examination of animals when anthrax is suspected. The disposal of animal carcases on farm can cause significant dangers. It may lead to contamination of soil and water. There is also a risk of spreading of disease to livestock on the farm or on neighbouring farms as well as risk to the public health. Therefore, correct disposal of carcases and infected/contaminated material are of considerable importance in preventing long distance and international transmission of anthrax. Carcases can be disposed of by following methods:

1. Burial Method

A deep (2.4 to 2.7 m) pit is constructed for burial of carcase. The carcase is lowered into the pit on to a thick layer of quicklime and then covered with enough quicklime and soil. Bury the carcase with a soil-coverage depth of at least 1 m, preferably 2 m. The buried carcases should be protected from stray or wild animals. Burial site can be protected either by applying strong smelling disinfectant or by barbed-wire fencing.

2. Burning or Incineration

Burning is the preferred way to dispose of infected carcasses. Ensure proper and complete burning from beneath. Incineration can be done either by cross-pit or by surface burning method. Cross-pit method is suitable for few carcases, while, surface burning method for a large number of carcases.

Cross-Pit Method

Two trenches are dug to form a cross with right angles. Each trench is 2 m long and 40 cm wide and 45 cm deep. Trenches are constructed deep at the cross and shallow at their ends. The soil is heaped at right angle-site to form four mounds.

These mounds are supported by iron or wooden bars. The tree branches or wooden logs are used to form the pyre. The carcase is put on pyre. Carcases and firewood are soaked with kerosene or paraffin and burnt.

Surface Burning Method

A trench of 90 cm long, 30 cm wide and 45 cm deep is dug for each carcase. The length varies with the number of animals to be cremated. Carcases are arranged in a row so that head of one carcase is next to the quarter of the adjacent carcase. The carcases are covered with dry grasses, straw and firewood soaked with kerosene or paraffin and burnt.

Precautions to be taken while Disposing the Carcases Died of Anthrax

1. Avoid opening of the carcase suspected of anthrax. In the unopened carcase, anthrax bacilli do not sporulate and are destroyed by the putrefactive process.

2. Carcases must be disposed of properly and carefully on an appropriate site.

3. Bedding and infected/contaminated materials should also be disposed of preferably by incineration.

4. Disposal site should be away from the water sources. Do not bury carcasses any closer than 250m from any drinking water supply; 50m from any watercourse or 10m from any filled drain.

Chapter 8

Bartonelloses

The genus *Bartonella* consists of several species of very small Gram-negative bacilli, mainly arthropod-borne, that are responsible for feverish illness in humans involving RBCs. *Bartonella* spp. causes several diseases in humans. The three most common are cat scratch disease, trench fever and Carrion's disease. Other diseases caused by bartonellae include bacillary angiomatosis, bacillary peliosis and subacute endocarditis. Bacillary angiomatosis is caused by *B. henselae* and *B. quintana*. Bacillary angiomatosis may present as lesions in the skin, subcutaneous tissue, bone, or other organs. Bacillary peliosis is caused by *B. henselae* and occurs primarily in immunocompromised people, such as those with advanced HIV infection. Bacillary peliosis causes vascular lesions in the liver and spleen. Many *Bartonella* species can cause subacute endocarditis, particularly infection of the heart valves.

Cat Scratch Disease (CSD)

Etiology

The disease is caused by *Bartonella henselae.* Bartonellae are small, slightly curved Gram-negative rods.

Epidemiology

Cat scratch disease occurs worldwide and may be present wherever cats are found. Stray cats may be more likely than pets to carry *Bartonella* spp. In the United States, most cases of CSD occur in the fall and winter. Ticks may carry some species of *Bartonella*, but there is currently no evidence that ticks can transmit *Bartonella* infection to humans. The disease occurs most frequently in children under age of 15 years.

Transmission

The disease in humans may be transmitted due to the scratches of domestic or feral cats, particularly kittens. Cats can harbor infected fleas that carry *Bartonella* organisms. These organisms can be transmitted from a cat to a person during a scratch. Some evidence suggests that CSD may be transmitted directly to humans by the bite of infected cat fleas.

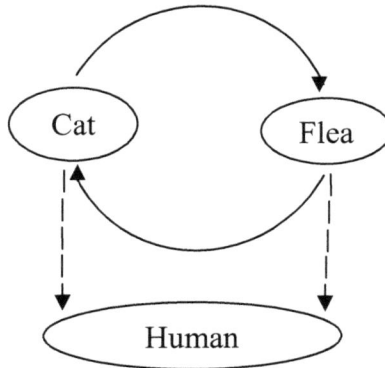

Figure 8.1: Infection Cycle of CSD.

Disease in Animals

Cat scratch disease does not occur in domestic animals. The reservoir of *B. henselae* is domestic cats, which do not show overt symptoms but can be bacteremic for long periods. *B. henselae* has also been detected in the fleas of infected cats.

Disease in Humans

The disease is characterized by mild fever, enlargement and tenderness of lymph nodes that develop 1-3 weeks after exposure of infection. There may be development of a papule or pustule at the site of inoculation. Rare manifestations such as eye infections, severe muscle pain, or encephalitis may be noticed.

Diagnosis

1. **Clinical examination:** Scratches, skin lesions, regional lymphadenopathy *etc.* may be suggestive of CSD.

2. **Histological examination:** There may be multiple abscesses with central necrosis surrounded by epithelioid cells, eosinophils and occasionally giant cells.

3. **Serological examination:** ELISA and IFA are important tests for diagnosis of CSD.

4. **Molecular diagnosis:** PCR technique is useful for confirmatory diagnosis of CSD.

Disease Management in Animals

1. Keep cats indoors and away from stray cats.
2. Use a flea collar or similar topical preventive.

3. Treat cats for fleas using fipronil (a broad spectrum insecticide that disrupt the insect's central nervous system) and other spot-on treatments.

Disease Management in Humans

1. Wash hands promptly after handling cats.
2. Immunocompromised individuals should avoid rough play with cats, particularly strays and kittens, to prevent scratches.
3. Immunocompromised individuals should avoid owning cats less than one year of age.
4. Avoid contact with flea infected cats.
5. Treatment with azithromycin.

Trench Fever

Synonyms

Five-days fever, Quintan disease

Trench fever received its name during World War I, when many soldiers fighting in the European trenches harbored infected body lice and became infected with the disease.

Etiology

The disease is caused by *Bartonella quintana*. This organism was formerly classified among the rickettsiae as *Rochalimaea quintana*. Unlike *Bartonella bacilliformis* the organism does not possess flagella, although may exhibit a twisting movement caused by fimbriae.

Epidemiology

Trench fever has a worldwide distribution. The cases have been reported from Europe, North America, Africa, and China.

Transmission

Trench fever is transmitted by the human body louse (*Pediculus humanus*). Because of its association with body louse infestations, trench fever is most commonly associated with homeless populations or areas of high population density and poor sanitation.

Disease in Humans

The disease is characterized by fever (which may be recurrent), headache, rash, papule on skin, lymphadenopathy, bone pain, mainly in the shins, neck and back.

Diagnosis

1. **Cultural examination:** The organism can be isolated from blood during all stages of infection using agar medium containing rabbit serum.

2. **Morphological examination:** Confirmatory diagnosis can be made by blood smear examination using Giemsa stain. They are seen packing the cytoplasm of the cell and adhering to the cell surface.

3. **Serological examination:** ELISA and counter-immuno-electrophoresis are important serological tests for diagnosis of trench fever.

4. **Molecular examination:** Confirmatory diagnosis of *Bartonella quintana* can be made using PCR amplification of the 16S rRNA or citrate synthetase genes.

Disease Management in Humans

1. Adopt proper personal hygiene.
2. Avoid exposure to human body lice. Body lice are typically associated with conditions of crowding.
3. Do not share clothing, beds, bedding *etc.* used by a person infested with body lice.

Carrion's Disease

Synonyms

Oroya fever, Verruga peruana

Carrion's disease, formerly known as bartonellosis, is transmitted by bites from sandflies (*Lutzomyia* spp.) that are infected with the organism.

Etiology

The disease is caused by *Bartonella bacilliformis*. In older cultures they tend to be extremely pleomorphic. They are slow growing on culture medium and may take up to 10 days.

Epidemiology

Carrion's disease has limited geographic distribution. Disease occurs in the Andes Mountains at 3,000 to 10,000 ft in elevation in western South America, including Peru, Colombia, and Ecuador. Most cases are reported in Peru. A few cases of Oroya fever and Verruga peruana (Peruvian warts) have been reported in travelers who returned from the Andean highlands in South America.

Transmission

Bartonella bacilliformis is pathogenic only to humans. The infection is spread by the sandfly *Lutzomyia verrucarum*. Individual may act as carriers of infection long after recovery from the illness and also after asymptomatic infection.

Disease in Humans

The disease occurs in two distinct phases:

1. Oroya Fever

During this phase, the clinical manifestations may be fever, headache, myalgia, abdominal pain, and severe anemia.

2. Verrugaperuana

During this later phase, lesions appear under the skin as nodular growths, and then emerge from the skin as red-to-purple vascular lesions that are prone to ulceration and bleeding.

Diagnosis

1. **Cultural examination:** Blood culture should be carried out at all stages of infection, on semisolid medium containing rabbit serum. It may be difficult to isolate the organisms from the blood when the verruga stage has developed and culture from the skin lesions is rarely satisfactory.

2. **Morphological examination:** The confirmatory diagnosis can be made by demonstration of the organisms in blood smears stained by Giemsa stain. They are seen packing in the cytoplasm of the cells and adhering to the cell surfaces.

Disease Management in Humans

1. **Control of sandflies:** Insecticide such as DDT is used to eliminate the sandfly vector in likely breeding sites inside and outside of houses and the surrounding areas. Since insects bite at only at night, individual may protect themselves by withdrawing from affected areas at nightfall. The disease can also be managed by using repellents and protective clothing to avoid sandfly bites in areas where Carrion's disease is common (South America). Avoid outdoor activities at dawn and dusk. During these times the sandflies are most active.

2. **Treatment:** Penicillin, streptomycin, tetracycline and chloramphenicol may be effective in the treatment of the infection and reducing the mortality rate. Blood transfusion may be necessary in severe cases of anemia.

Chapter 9

Borrelioses

Borreliosis is an important zoonotic disease. Lyme borreliosis and relapsing fever are important from public health point of view. There are several species of *Borrelia* occurring in animals have not been described as infectious agents in humans. *Borrelia anserine* is found worldwide in geese, ducks, chickens and turkeys.

Lyme Borreliosis

Synonyms

Lyme disease, Lyme arthritis

Lyme borreliosis or Lyme disease, originally called Lyme arthritis, was recognized as an infectious condition in 1975, following an epidemiological investigation of a cluster of cases of suspected juvenile rheumatoid arthritis which occurred in Lyme, Connecticut, USA. Lyme borreliosis shows various manifestations in animals and humans.

Etiology

The disease is caused by *Borrelia burgdorferi* sensu lato which is a Gram-negative and motile organism.

Epidemiology

The disease is worldwide in distribution. The reservoirs of *Borrelia burgdorferi* are wild and domesticated animals including mice and other rodents, cattle, sheep, deer, horses and dogs. Infections occur more commonly during summer or early autumn due to increased activity of tick vector during these seasons.

Transmission

The disease is transmitted by infected tick (*Ixodes* spp.) bite. Borreliae manage to get into the bite wound through infected tick saliva.

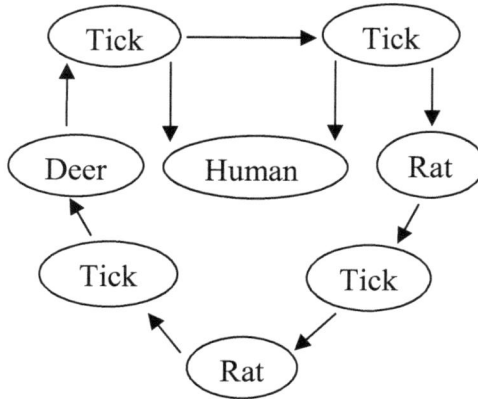

Figure 9.1: Infection Cycle of Lyme Borreliosis.

Disease in Animals

Lyme borreliosis may be inapparent, though clinical infection has been observed in cattle, horses and dogs. The disease is characterized by fever, loss of condition, nervous symptoms, arthritis, arthralgia and muscular tenderness. It is one of the most common tick-transmitted diseases in the world but only causes symptoms in 5-10 per cent of affected dogs.

Disease in Humans

The disease is characterized by erythema migrans on skin, regional lymphadenopathy *etc*. Systemic infection may involve myalgia, arthritis, arthralgia, severe headache and CNS symptoms.

Diagnosis

1. **Clinical examination:** The disease can be diagnosed by history of tick bite and clinical symptoms.

2. **Cultural examination:** The organism can be isolated by using Barbour-Stoenner-Kelly (BSK) medium. Skin biopsy is more suitable for isolation of the organism. However, it can also be isolated from blood and CSF, but recovery is less.

3. **Serological examination:** The disease can be diagnosed by using Immunofluorescence assay and ELISA. CDC currently recommends a two-step process when testing blood for evidence of antibodies against the Lyme disease bacteria. Both steps can be done using the same blood sample. The first step uses a testing procedure called "EIA" (enzyme immunoassay) or rarely, an "IFA" (indirect immunofluorescence assay). If this first step is negative, no further testing of the specimen is recommended. If the first step is positive or indeterminate (sometimes called "equivocal") the second step should be performed.

4. **Molecular examination:** Confirmatory diagnosis of the disease can be made by using PCR technique.

Disease Management in Animals

Dogs should not be allowed to roam in tick-infested environments where Lyme disease is common. Ticks from the body of the dogs should be removed. Sprays, collars and spot-on topical products can be applied to repel and kill the ticks. Lyme vaccines are available, but their use is somewhat controversial. The disease can be treated by using antibiotics.

Disease Management in Humans

1. Repellents, insecticide dips and environmental sanitation can be applied for protection from ticks.
2. Vaccination can be done by using whole cell bacterins.
3. Prophylaxis with penicillin and treatment with doxycycline are effective.

Relapsing Fever

Relapsing fever occurs worldwide in endemic or epidemic forms. Both are transmitted by arthropod vectors. The disease is characterized by recurrence of fever and spirochaetaemia.

Etiology

Endemic or tick-borne relapsing fever is caused by *Borrelia duttonii*, while epidemic or louse-borne relapsing fever is caused by *Borrelia recurrentis*.

Epidemiology

Tick-borne relapsing fever is endemic in Africa. Wild rodents and domestic animals serve as reservoirs. The louse-borne form is endemic in Ethiopia, Sudan, Somalia, Chad, Bolivia and Peru.

Transmission

The disease is transmitted in humans by human body louse (*Pediculus humanus corporis*) and tick (*Ornithodoros* spp.). Infection in humans does not occur through louse bites but rather through infected hemolymph, which is released when lice are crushed. The organism is capable of penetrating intact skin. Tick-borne relapsing fever is transmitted to humans through infected saliva or coxal fluid during tick bite.

Disease in Humans

The disease is characterized by periodic bouts of fever, nausea, vomiting, splenomegaly, hepatomegaly *etc.*

Diagnosis

1. **Microscopic examination:** Organisms can be examined in blood smear from a febrile stage of disease.

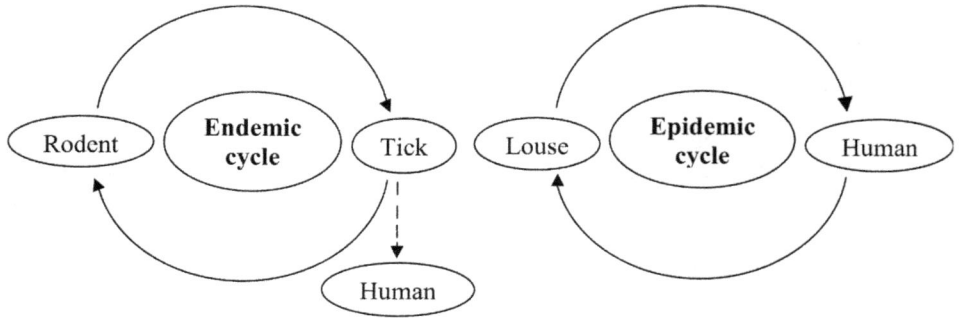

Figure 9.2: Infection Cycle of Relapsing Fever.

2. **Serological diagnosis:** It can be done by using ELISA and IFA, but these tests may give false positive diagnosis.

3. **Molecular diagnosis:** PCR technique has been developed for diagnosis of this disease.

Disease Management in Humans

1. Control of lice and ticks.
2. Protection from ticks and lice.
3. Treatment with doxycycline or erythromycin.

Chapter 10

Brucellosis

Synonyms

In humans- Undulant fever, Mediterranean fever, Malta fever, Gibraltar fever, Cyprus fever; In animals- Contagious abortion, Infectious abortion, Enzootic abortion, Bang's disease.

Brucellosis is a bacterial zoonotic disease caused by various *Brucella* species, which mainly infect cattle, sheep, goat, swine and dogs. Humans generally acquire the disease through direct contact with infected animals, by eating or drinking contaminated animal products, or by inhaling airborne agents. The majority of cases are caused by ingesting unpasteurized milk or cheese from infected goats or sheep. The disease is characterized fever, weakness, malaise and weight loss.

Etiology

B. abortus causes most brucellosis in cattle, but *B. melitensis* and *B. suis* can also cause bovine infection. *B. melitensis* is the main cause of brucellosis in sheep and goats and *B. suis* in swine.

Epidemiology

The important animal species are cattle, sheep, goats, pigs that serve as source of infection. Less important animal species are bison, buffalo, camels, dogs, horses, reindeer and yaks, but they can be very significant local sources of infection in some regions. Recently, the infection has also been reported in marine mammals, including dolphins, porpoises and seals and these animals may present an emerging hazard to persons occupationally exposed to infected tissues from them.

B. melitensis is most frequently reported as a cause of human disease and the most frequently isolated from cases. It is the most virulent type and associated with severe acute disease. It is recorded as endemic in several countries and accounts for a disproportionate amount of human brucellosis. The organism primarily infects sheep and goats, but other animal species, including dogs, cattle and camels can also be infected. *B. abortus* causes the most widespread infection, but associated with much less human disease. Generally, the infection in humans is sub-clinical and cause less severe disease than that caused by *B. melitensis* or *B. suis*. Cattle are the most common source of *B. abortus* but bison, buffalo, camels, dogs and yaks are important in some regions. *B. suis* has restricted occurrence than *B. melitensis* and *B. abortus*. It is locally important as a source of human infection which can be as severe as that produced by *B. melitensis*. *B. canis* cause widespread infection of dogs in many countries. It is less frequently associated with human disease and cause mild disease. *Brucella* infection also occurs in many wild animal species but these are rarely implicated as sources of human infection.

Brucellosis is an important human disease in many parts of the world especially in the Mediterranean countries of Europe, North and East Africa, the Middle East, South and Central Asia and Central and South America and yet it is often unrecognized and frequently goes unreported. There are only a few countries in the world that are officially free of the disease although cases still occur in people returning from endemic countries. Expansions of international travel which stimulates the taste for exotic dairy goods such as fresh cheeses which may be contaminated, and the importation of such foods into *Brucella*-free regions, also contribute to the ever-increasing concern over human brucellosis. Brucellosis affects people of all age groups and of both sexes. Brucellosis causes more than 500 000 infections in humans worldwide every year.

Transmission

Cattle, sheep, goats and pigs are the main reservoirs of *Brucella* spp. Transmission occurs by direct contact and environmental contamination following abortion. Sexual transmission and/or artificial insemination are also important. Seronegative latent infections can occur. Transmission to humans occurs through occupational or environmental contact with infected animals or their products. Foodborne transmission is a major source of infection, with cheese made from raw milk and unpasteurized milk presenting a high risk. Brucellosis can be a travel-associated disease. Blood or organ/tissue transfer is possible sources of infection. Person-to-person transmission is extremely rare.

Disease in Animals

Brucellosis infects many species, especially cattle, sheep, goats, pigs. Different *Brucella* types infect different species preferentially. Brucellosis presents typically as abortion in animals.

Disease in Humans

Human brucellosis usually presents as an acute febrile illness. Most cases are caused by *B. melitensis*. All age groups are affected. Childhood brucellosis once considered rare in children, it is now recognized that brucellosis can affect persons of all ages, especially in areas where *B. melitensis* is the predominant species. The

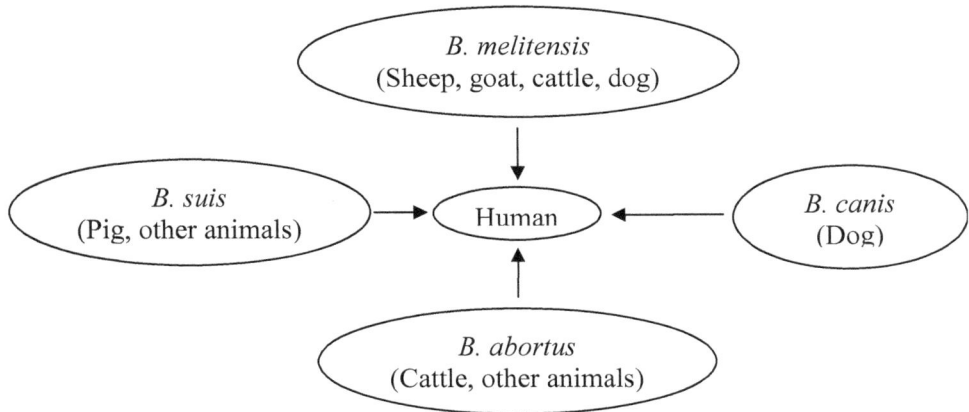

Figure 10.1: Infection Cycle of Brucellosis.

course of infection and the incidence of complications appear to be similar regardless of the age of the patients. The disease may persist as relapse, chronic localized infection or delayed convalescence. Complications may affect any organ system.

Gastrointestinal complications occur due to foodborne brucellosis caused by ingestion of unpasteurised milk and milk products from animals infected with *B. Melitensis*. Foodborne brucellosis resembles typhoid fever, in that systemic symptoms predominate over gastrointestinal complaints. Associated symptoms are nausea, vomiting, and abdominal discomfort. Hepatobiliary complications are caused by *B. abortus* may show epithelioid granulomas. *B. Melitensis* causes scattered small foci of inflammation resembling viral hepatitis.

Osteoarticular complications are the most frequent complications of brucellosis, occurring in up to 40 per cent of cases. A variety of syndromes have been reported, including sacroiliitis, spondylitis, peripheral arthritis, osteomyelitis, bursitis, and tenosynovitis. Respiratory tract complications include hilar and paratracheal lymphadenopathy, interstitial pneumonitis, bronchopneumonia, lung nodules, pleural effusions and empyema. Cardiovascular complications involve endocarditis which is the most common cause of death from brucellosis. Neurological complications (neurobrucellosis) occur due to direct invasion of the central nervous system with *B. Melitensis* which has been reported in 5 per cent cases. Cutaneous complications include rashes, nodules, papules, erythema nodosum, petechiae, and purpura. Ophthalmic complications include uveitis which is the most frequent manifestation and can present as chronic iridocyclitis, nummular keratitis, multifocal choroiditis or optic neuritis.

Genitourinary complications are the most frequent complications of brucellosis in men which include orchitis and epididymitis. Brucellosis during the course of pregnancy carries the risk of spontaneous abortion or intrauterine transmission to the infant. Abortion is a frequent complication of brucellosis in animals, where placental localization is believed to be associated with erythritol, a growth stimulant for *B.*

abortus. Although erythritol is not present in human placental tissue, bacteremia can result in abortion, especially during the early trimesters.

Diagnosis in Animals

Culture of *Brucella* from abortion material, milk or tissues collected at autopsy provides a definitive diagnosis. Farrell's medium (serum dextrose agar with antibiotics and other drugs) is a selective medium for isolation of brucellae. Media containing bacteriostatic dyes, which are inhibitory to strains of *B. abortus* biotype 2 and other fastidious strains. Antibiotics used in place of dyes enabled growth of all biotypes *Brucella* spp. to appear on selective media. This medium is incorporated with the bacitracin 25 IU/ml, vancomycin 20 mg/ml, cycloheximide 100 mg/ml and nystatin 100 IU/ml in a serum-glucose agar base. On serum dextrose agar without dyes, the colonies are pale yellow and translucent. Serology is usually the most practicable method. RBT is recommended for screening in cattle; ELISA or complement fixation is recommended for confirmation of infection in individual animals. Screening of milk samples by milk ring test or ELISA is useful for surveillance. No single serological test is reliable for confirmation of infection in sheep, goats and pigs. Serological tests should be used on a herd or flock basis. Similarly, the skin test is useful for screening at the herd or flock level, especially if vaccination is not used.

Diagnosis in Humans

The disease can be insidious and may present in many atypical forms. In many patients the symptoms are mild and, therefore, the diagnosis may not be even considered. Indeed it should be noted that even in severe infections differential diagnosis can still be difficult. The application of well-controlled laboratory procedures and their careful interpretation can assist greatly in this process. In acute brucellosis, isolation of *Brucella* from blood or other tissues is definitive. Culture is often negative, especially in long-standing disease. Serology is generally useful diagnostic approach. The RBT, tube agglutination and ELISA procedures are recommended. Methods which differentiate IgM and IgG can distinguish active and past infection. False positive serological reactions may occur. Skin test reactions indicate past exposure not active infection.

Disease Management in Animals

1. **Hygienic and managemental practices:** Animal brucellosis is best prevented by careful herd management and hygiene.

2. **Vaccination:** *B. abortus* strains 19 and RB 51 are recommended for prevention of bovine brucellosis. *B. melitensis* Rev 1 is recommended for prevention of *B. Melitensis* infection in sheep and goats.

3. **Health education:** Education and information programmes are essential to ensure cooperation at all levels in the community.

4. **Eradication:** Eradication can only be achieved by test-and slaughter combined with effective prevention measures and control of animal movements.

Disease Management in Humans

1. **Vaccination:** Vaccination is not generally recommended.

2. **Treatment:** It requires the administration of effective antibiotics for an adequate length of time. Treatment of uncomplicated cases in adults and children 8 years of age and older is (doxycycline 100 mg twice a day for six weeks + streptomycin 1g daily for 2-3 weeks) or (doxycycline 100mg twice a day for six weeks + rifampicin 600-900mg daily for six weeks).

3. **Heat treatment of milk and meat and their products:** All dairy products should be prepared from heat-treated milk. Consumption of raw milk or products made from raw milk should be avoided. Meat should be adequately cooked.

4. **Health education:** Public health education should emphasize food hygiene and occupational hygiene.

5. **Hygienic measures:** The prevention of human brucellosis is based on occupational hygiene and food hygiene.

6. **Miscellaneous measures:** Special precautions should be taken by laboratory workers. Physicians and health workers should be aware of the possibility of brucellosis.

Surveillance of Brucellosis in Humans and in Animals

Continued surveillance is essential to monitor the presence/absence of brucellosis and the efficacy of control programmes. The key to effective surveillance is the case definition, reporting, analysis of data and dissemination of information for action. The surveillance programme must be designed according to the adopted control strategy. Human cases may be the first indication of infection in the animal population.

Chapter 11

Chlamydial Infections

Chlamydiae are small, obligatory intracellular bacteria which undergo a characteristic developmental cycle. Chlamydiae belong to the family Chlamydiaceae. This family has two important genera, *viz.*, *Chlamydia* and *Chlamydophila*. The zoonotic species of Chlamydiae belongs to the genus *Chlamydophila*, *e.g.*, *Chlamydophila psittaci*, *Chlamydophila abortus* and *Chlamydophila felis* cause zoonosis in avian, ruminants and cats, respectively.

Psittacosis/Ornithosis

This is an important chlamydial zoonotic disease transmitted in human beings from psittacine birds. The disease is characterized by fever, chills, headache, myalgia, photophobia, interstitial pneumonia and cough.

Etiology

The disease is caused by *Chlamydophila psittaci.* This is an obligatory intracellular organism. The organisms divide by binary fission.

Epidemiology

The disease is worldwide in distribution. The natural reservoirs are psittacine birds. The pigeon breeders and poultry employees are at greater risk of acquiring infections. Asymptomatic birds, particularly psittacine birds and pigeons are the most important source of infection.

Transmission

1. **Inhalation:** The disease is transmitted to human beings by inhalation of contaminated dust.
2. **Contact:** The disease can also be transmitted through contact with excretions of infected birds.

Disease in Humans

The incubation period ranges 1-3 weeks which may extend to 3 months. Clinical manifestations are quite variable from inapparent infections to severe life-threatening pneumonia with high fever, severe headache and multi-organ failure. Other symptoms are bradycardia, myalgia, malaise and cough.

Diagnosis

The disease can be diagnosed by using ELISA; however, the PCR technique is highly sensitive.

Disease Management in Humans

1. **Prophylaxis:** Persons at risk *i.e.* pigeon breeders and employees in poultry industry should wear protective clothing and face masks. Regular serological monitoring of persons at risks should be carried out.

2. **Treatment:** The drug of choice of is doxycycline (100 mg, orally, twice a day). Alternative drugs such as tetracycline (500 mg, orally, four times a day) or azithromycin (500 mg on first day followed by 250 mg daily) can be used for treatment. The treatment is continued for 2-3 weeks.

Chapter 12

Ehrlichioses

Ehrlichiosis is a vector-borne bacterial zoonotic disease. The disease is found in acute or chronic form. The acute form of disease is characterized by nausea, headache, myalgia, arthralgia and liver and kidney damage.

Etiology

The disease is caused by *Ehrlichia* spp. The organism is non-motile, pleomorphic and obligatory intracellular Gram-negative bacteria.

Epidemiology

The disease has been reported from US, UK, Spain, Portugal, Italy, Mali (Africa), Japan and Malaysia. *E. canis, E. platys* and *Neorickettsia helminthoeca* have been found in dogs. *E. canis* causes tropical canine pancytopenia. *E. platys* causes infectious granulocytic ehrlichiosis. *E. equi* causes equine ehrlichiosis. *E. phagocytophila* causes sheep tick fever. Human ehrlichiosis is caused by *E. chaffeensis, E. canis, E. ewingii, E. equi* and *E. senntsu*. The prevalence of disease is higher in June and July months (highest tick activity).

Transmission

The disease is transmitted by infected tick bite. The important species of ticks, *viz., Dermacentor* spp., *Ixodes* spp. and *Rhipicephalus* spp. have been involved in the transmission of ehrlichiosis.

Disease in Humans

Ehrlichiae affect multi organs of the body. Organisms cause lymphocytic infiltration in the stomach, intestine, heart, kidney and meninges. They also cause granulomas and necroses in liver, lungs, spleen, bone marrow and lymph nodes. Following an incubation period of 9 days, the disease is manifested by fever, chills, headache, nausea, vomiting, malaise, myalgia and arthralgia. In severe cases there is involvement of central nervous system which is characterized by confusion, polyneuropathy, meningeal irritations *etc.* Cutaneous, pulmonary and intestinal hemorrhages also occur in severe cases.

Diagnosis

1. **Clinical examination:** Clinical symptoms like high fever, chills, headache, nausea, vomiting, malaise, myalgia, arthralgia, granulocytopenia and thrombocytopenia are indicative of ehrlichiosis.

2. **Immuno-histological examination:** This method can be used to detect *Ehrlichia* spp. in lymphocytes, biopsy materials and bone marrow.

3. **Serological examination:** Immunofluorescence assay and ELISA have been useful in the diagnosis of ehrlichiosis.

4. **PCR technique:** Species identification of *Ehrlichia* can be done by using PCR technique.

Disease Management in Humans

1. **Control of ticks:** Tick contact should be avoided by using protective clothing and insect repellents.

2. **Treatment:** The drug of choice is doxycycline (100 mg, orally, twice daily for 2 weeks). Chloramphenicol (500 mg, orally, four times a day for 2 weeks) is an alternative drug.

Chapter 13

Erysipeloid

Synonym

Swine erysipelas

Erysipeloid is an acute infection of pig and other animals including humans. The disease is characterized by skin lesions with rare complications of septicemia and endocarditis.

Etiology

The disease is caused by *Erysipelothrix rhusiopathiae* which is a non-motile, non-spore forming Gram-positive rod.

Epidemiology

The disease is worldwide in distribution. It is found primarily in soil and decomposing vegetation. The persons at risks are butchers, veterinarians, farmers, animal keepers and employees of poultry and meat industry. Pig, poultry and fish are important source for human infection.

Transmission

1. **Contact:** The infection is transmitted in humans through contact with pigs, poultry particularly turkey and ducks and fish. The infection in person is transmitted though cuts, stabs and tears during handling of infected animal tissues or contaminated instruments.

2. **Ingestion:** The infection is rarely transmitted through consumption of infected meat. In case of swine, consumption of fish meal is the main source of infection. Infection in animals can also be transmitted by contaminated drinking water, soil and bedding materials.

3. **Bite:** The infection is rarely transmitted through dog bites.

Disease in Animals

In pigs, the disease may occur either in acute or chronic form. Acute form is characterized by septicemia, while chronic form by cutaneous erysipelas, arthritis and endocarditis. In sheep, particularly young ones, the disease occurs in chronic form which is characterized by polyarthritis, while acute septicemic disease is rare. In poultry and wild birds, the disease is characterized by fever, dyspnea and diarrhea.

Disease in Human

Following an incubation period of 2-5 days, there is development of inflammation at the site of inoculation mostly on the hands and fingers. Initially the skin lesion is red but later on become bluish and then blanches from the center towards periphery. The affected skin area is slightly swollen, itches and become painful. There is no fever or pus formation in the affected area of skin. Lymphangitis and arthritis are rare. The complication of disease may lead to septicemia and endocarditis.

Diagnosis

1. **Cultural examination:** Confirmatory diagnosis of erysipelas can be made by cultural examination of organism. The suitable material for this purpose can be taken from the margins of the erythema by skin biopsy. In systemic illness, the suitable clinical material is blood. Serological tests are not useful for diagnosis of erysipeloid.

2. **PCR technique:** This technique is useful for confirmatory diagnosis of erysipelas.

Disease Management in Animals

1. Avoid feeding of contaminated feed and water.
2. Isolation of infected animals from healthy stock.
3. Vaccination with live attenuated vaccines.
4. Acute disease is treated with erysipeloid antiserum and penicillin.

Disease Management in Humans

1. **Prophylaxis:** Direct contact with infected animal tissues or materials such as carcase, fur, bones *etc.* should be avoided by using gloves.

2. **Treatment:** Penicillin V can be given at the dose rate of 1 g, orally, daily for 1 week to shorten the course of disease. Penicillin G (10-20 million IU, daily, i.v. for 4-6weeks) is effective in septicemia. Erythromycin can be given as an alternative drug in penicillin allergic patients.

Chapter 14

Glanders

Synonyms

Malleus, Farcy, Morve, Muermo

Glanders is a highly fatal bacterial zoonotic disease that primarily affects horses, mules and donkeys. Some animals die acutely within a few weeks. Others become chronically infected, and can spread the disease for years before succumbing. Glanders also occurs occasionally in other mammals, including carnivores that eat meat from infected animals. Although cases in humans are uncommon, they can be life threatening and painful.

Etiology

Glanders is caused by *Burkholderia mallei*, a Gram-negative rod which belongs to the family Burkholderiaceae. The organism was formerly known as *Pseudomonas mallei*.

Epidemiology

The main reservoirs of *B. mallei* are horses, mules and donkeys. Most other domesticated mammals can be infected experimentally (pigs and cattle are reported to be resistant), and naturally occurring clinical cases have been reported in some species. Members of the cat family seem to be particularly susceptible, with cases documented in domesticated cats, tigers, lions, leopards and other felids. Deaths have also been reported in other carnivores that ate glanderous meat, including dogs, bears, wolves, jackals and hyenas. Birds are highly resistant to *B. mallei*.

Glanders is thought to be endemic in parts of the Middle East, Asia, Africa and Central and South America. This disease has sometimes re-emerged in countries where it appeared to be absent or was limited to small foci of infection (*e.g.*, India in 2006). It has been eradicated from Western Europe, Canada, the US, Australia,

Japan and some other countries. The geographic distribution of *B. mallei* can be difficult to determine precisely, as cross-reactions with *B. pseudomallei* interfere with serological surveys.

Transmission

Glanders is mainly transmitted by contact with infected horses, mules and donkeys. The transmission occurs most oftenly via their respiratory secretions and exudates from skin lesions. Chronically or subclinically infected equids can shed *B. mallei* intermittently or constantly. The organism can enter the body by contamination of abraded skin and mucous membranes, or inhalation of aerosols. Equines often become infected due to ingestion of food or water contaminated with *B. mallei.* Carnivores become infected due to ingestion of contaminated meat. Venereal transmission from stallions to mares, and vertical transmission from the dam has also been reported. *B. mallei* is readily spread on fomites including harnesses, grooming tools, and food and water troughs. Flies might act as mechanical vectors.

In humans, the disease is transmitted through contact with sick animals, contaminated fomites, tissues and bacterial cultures. The organism enters the body through wounds and abrasions in the skin, and by ingestion or inhalation. Laboratory-acquired infections occur during routine handling and processing of cultures or samples, rather than after injuries or accidents. Person-to-person transmission of infection has been rarely reported.

Disease in Animals

The incubation period in equids is 2-6 weeks. Infections can be latent for varying periods. In equids, glanders is traditionally categorized into nasal, pulmonary and cutaneous forms, based on the most commonly affected sites. In the nasal form, deep ulcers and nodules develop inside the nasal passages, resulting in a thick, mucopurulent, sticky, yellowish discharge. This discharge can be copious, may be unilateral or bilateral, and can become bloody. The ulcers may coalesce over wide areas, and nasal perforation is possible. Healed ulcers become star-shaped scars, which may be found concurrently with nodules and ulcers. The submaxillary lymph nodes become enlarged, indurated and occasionally suppurate and drain. Pulmonary involvement occurs in combination with other forms of glanders. This form is characterized by development of nodules and abscess in lungs, bronchopneumonia, cough, dyspnea, fever and progressive debilitation.

Cutaneous form of ganders (farcy) is characterized by development of multiple nodules in the skin along with the course of lymphatic vessels, which often rupture, ulcerate and discharge an oily, thick yellow exudate. Glanders ulcers heal very slowly, often continuing to discharge fluid, although dry ulcers may also be seen. The regional lymphatics and lymph nodes become chronically enlarged, and the lymphatics are filled with a purulent exudate. Other symptoms such as swelling of the joints, painful edema of the legs or glanderous orchitis may be seen in some animals. The skin lesions are most commonly found on the inner thighs, limbs and abdomen. In cutaneous form of glanders, the animals can remain in good condition for a time, but they eventually become debilitated and die.

Disease in Humans

The incubation period may vary from 1 to 14 days (generally 5 days). In chronic cases, the incubation period may extend to months. Glanders occurs in different forms such as septicemia, pulmonary infection, acute localized infection and chronic disease. One form can progress to another, and combinations of syndromes occur.

Localized infections are characterized by development of nodules, abscesses and/or ulcers in the mucous membranes, skin and/or subcutaneous tissues at the site of inoculation. Initially the skin lesion may appear as a blister that gradually develops into an ulcer. Involvement of the lymphatics in the area results in lymphangitis with numerous foci of suppuration. Involvement of mucous membranes may lead to mucopurulent, sometimes blood-tinged discharge. There may be swelling of nose and the face due to local tissue destruction in the nasal passages. Localized lesions may be accompanied by systemic signs of illness, including fever, sweats, malaise, headache and swelling of the regional lymph nodes. Mucosal or skin infections may disseminate to other organs such as lungs, spleen, liver and muscles. In disseminated cases, the nonspecific clinical signs such as nausea, dizziness, night sweats, myalgia, severe headache, papular or pustular rash, weight loss *etc.* may be seen. Disseminated infections often progress to septicemia.

The pulmonary form can occur acutely after inhaling *B. mallei*, but the organisms may also reach the lungs by localized or hematogenous spread from other forms. Pulmonary form is characterized by development of abscesses, pleural effusion and pneumonia, cough, chest pain, dyspnea, fever, chills, sweats, headache and myalgia. Lymphangitis, nasal and gastrointestinal signs may also be seen. The skin abscesses can develop up to several months after organisms are inhaled. Untreated pulmonary disease often develops into septicemia.

Septicemic form is characterized by acute onset of fever, chills, myalgia, headache and pleuritic chest pain. Other symptoms may include flushing, a pustular or papular rash, lymphadenopathy, cellulitis, cyanosis, photophobia, jaundice, diarrhea and granulomatous or necrotizing lesions, hepatomegaly and splenomegaly. Multi-organ failure is common, and death can occur rapidly.

Chronic glanders is characterized by multiple abscesses, nodules and ulcers in various tissues, with periodic recrudescence and milder symptoms than in acute cases. Weight loss, lymphadenopathy and lymphangitis are common.

In untreated disease, the case fatality may be very high as 95 per cent or more in septicemic form and 90-95 per cent in pulmonary form. With treatment, the case fatality rates for these forms are reported to be as high as 40-50 per cent. The case fatality rate in localized disease is 20 per cent.

Diagnosis in Animals

1. **Cultural examination:** Glanders can be diagnosed by culturing *B. mallei* from lesions, lymph nodes, and nasal or other respiratory exudates. This organism is uncommonly detected in blood. Bacteriological diagnosis can

be difficult when the animal is in the early stages of disease or subclinically infected.

2. **Serological examination:** CFT and ELISAs are currently considered to be the most accurate and reliable assays in equids. Cross-reactivity can be an issue in serological tests. Most serological tests cannot distinguish whether the animal has antibodies to *B. mallei* or *B. pseudomallei*.

3. **Mallein test (hypersensitivity reaction):** It was used in glanders eradication programmes, and is still used to detect infected equids in some countries. In the 3 versions of the mallein test, a protein fraction of *B. mallei* is injected into the eyelid (intradermo-palpebral test), administered in eyedrops, or injected subcutaneously at a site other than the eye. The intradermo-palpebral test is considered to be the most reliable and sensitive version. Reactors in this test develop marked eyelid swelling after 1 to 2 days. There may also be a purulent ocular discharge and an elevated body temperature. Conjunctivitis occurs after administration in eyedrops, and a firm, painful swelling with raised edges is seen within 24 hours after subcutaneous (non-ocular) injection. Mallein testing can cause transient false positives in subsequent serological tests.

4. **Molecular diagnosis:** Molecular techniques such as PCR, restriction fragment length polymorphism (RFLP), pulse-field gel electrophoresis (PFGE) and 16S rRNA sequencing can distinguish *B. mallei* from *B. pseudomallei*.

Diagnosis in Humans

1. **Cultural examination:** Glanders can be diagnosed by culturing *B. mallei* from lesions, as in animals. This organism may also be found in sputum, blood or urine, although blood cultures are often negative.

2. **Serological examination:** Serology can be employed in diagnosis, if tests are available; however, there can be unexplained high background titers in some normal sera, and seroconversion tends to occur late. Many serological tests cannot distinguish reactions to other species of *Burkholderia*, including *B. pseudomallei*. The mallein test is not used in humans.

3. **Radiological examination:** Radiography is helpful in the pulmonary form, although it is not specific for glanders. The lesions can include bilateral bronchopneumonia, miliary nodules, segmental or lobar infiltrates, and cavitating lesions. Similar lesions may also be detected in other organs and tissues.

4. **Molecular diagnosis:** PCR assays or antigen detection tests could be useful, though they are not employed routinely.

Disease Management in Animals

1. **Disease reporting:** A quick response is vital for containing outbreaks in glanders-free regions. Veterinarians who encounter or suspect this disease should follow their national and/or local guidelines for disease reporting.

2. **Prevention:** During outbreaks in non-endemic regions, animals that test positive are usually euthanized, and the premises are quarantined, cleaned and disinfected. Carcases, contaminated bedding and food should be safely destroyed (*e.g.*, burned or buried), and equipment and other fomites should be disinfected. In endemic areas, susceptible animals should be kept away from communal feeding and watering areas, since glanders is more common where animals congregate. Meat from infected equids should not be fed to other animals (or used for human consumption). Vaccines against glanders are not available.

3. **Treatment:** Some antibiotics may be effective against glanders, but treatment is often not allowed outside endemic areas. Treatment can be risky because treated animals may become asymptomatic carriers.

Disease Management in Humans

1. **Treatment:** Glanders is treated with antibiotics. Long-term treatment and/or multiple drugs may be necessary in some cases. Abscesses may need to be drained.

2. **Hygienic measures:** Strict precautions, including appropriate personal protective equipment (PPE), should be taken when handling infected animals and contaminated fomites.

3. **Miscellaneous measures:** Biosafety level 3 practices are required for manipulating infected tissues and cultures. No vaccine against glanders is available. Although person-to-person transmission is rare, human glanders patients should be isolated. Infection control precautions should be taken, and PPE, including disposable surgical masks, face shields, and gowns, should be used as appropriate during nursing.

Chapter 15

Leptospirosis

Synonyms

Weil's disease, Swine handler's disease, Rice field worker's disease, Sugarcane worker's disease, Mud fever, Swamp fever, Canicola fever, Seven day fever, Infectious jaundice, Canine typhus.

Leptospirosis is an acute or chronic contagious disease of domesticated and wild animals as well as humans. The disease is characterized by fever, anaemia, haemoglobinuria, icterus and abortion.

Etiology

The disease is caused by *Leptospira interrogans* which is pathogenic for animals as well as humans. Pathogenic leptospirae belong to the genus *Leptospira*, which is a long corkscrew-shaped bacterium.There are many host adopted serovars of *Leptospira interrogans viz.*, Hardjo, Pomona, Canicola and Icterohemorrhagiae in cattle; pig and cattle; dog and rat, respectively. The main source of infection is the urine of infected animals. Leptospirae are secreted in the urine of infected animals for a long time, often for an entire life time in case of rodents. Many wild and domestic animal species such as rats, mice, voles, cattle, buffalo, sheep, goat, pig and horse are reservoirs of this organism. Leptospirae shed in the urine of the reservoirs and can survive in the soil and water for many weeks. Therefore, environment may be highly contaminated where reservoir animals frequently urinate. Domestic animals may be infected through grazing in areas contaminated with urine of infected animals. Human beings usually get the infection from domestic animals.

Epidemiology

Leptospirosis is a zoonosis of worldwide distribution. The disease is endemic mainly in countries with humid subtropical or tropical climates and has epidemic potential. Seasonal outbreaks have been reported in northern Thailand and in Gujarat, India following heavy rainfall and flooding. Feral and domestic animals are the reservoir of leptospirae. Human infections occur accidentally. Children acquire the infection from dog more frequently than the adults. Human infections occur usually due to occupational exposure to the urine of the infected animals. Therefore, it is an important occupational zoonotic disease associated with farmers, livestock owners, veterinarians, rice field workers, sugarcane field workers and persons working in underground sewers. The outbreaks of disease mostly occur as a result of heavy rainfall or flood. In India, many outbreaks of leptospirosis have been reported from time to time. The outbreak of leptospirosis occurred in Odisha in October, 1999 after a super cyclone. Outbreaks of leptospirosis have also occurred in Gujarat, Kerala, Maharashtra and Andaman and Nicobar in August, 2000. In South-East Asia, leptospirosis outbreaks in humans are increasingly being reported during the rainy season in India, Indonesia, Sri Lanka and Thailand. Major outbreaks in the Region have been reported in Jakarta, Indonesia (2003), Mumbai, India (2005) and Kurunegala, Sri Lanka (2008).

Transmission

1. **Direct contact:** *Leptospira* can enter the body through skin abrasion or through intact mucous membrane by direct contact with urine or tissues of infected animal.

2. **Indirect contact:** Leptospirosis can be transmitted through contact of the abraded skin with soil, water or vegetation contaminated with urine of infected animals.

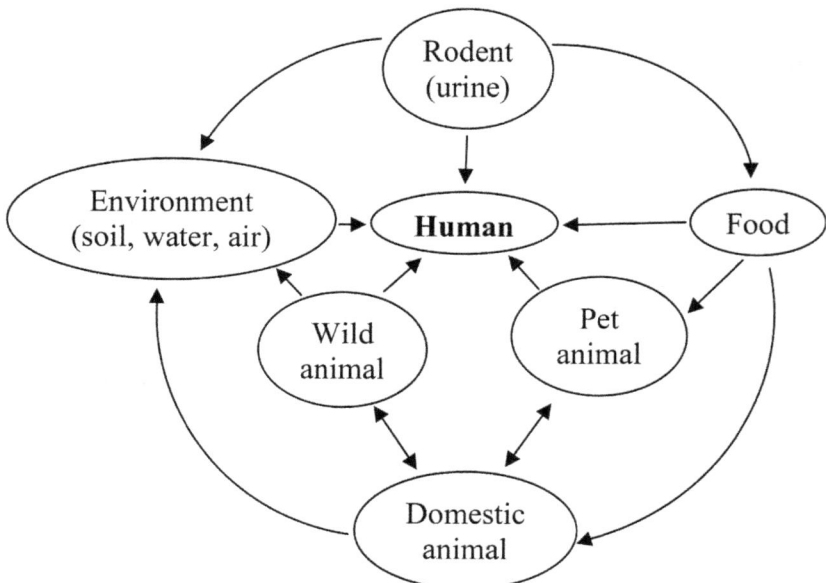

Figure 15.1: Infection Cycle of Leptospirosis.

3. **Ingestion:** Leptospirosis can be transmitted through ingestion of food or water contaminated with *Leptospira*.

4. **Inhalation:** Leptospirosis may also be transmitted through inhalation of droplets of urine during milking from infected cow or goat.

Note: Man does not transmit *Leptospira* to other persons and is therefore, called dead-end host.

Disease in Animals

In animals, the disease is found in acute, subacute or chronic form. Acute form is manifested by sudden onset of fever accompanied with dyspnea, haemoglobinuria, icterus and abortion in cattle and buffaloes. Cattle carcase affected with leptospirosis show icteric condition and multiple widespread of hemorrhagic spots below the mucus and serous membranes. Liver cells shows necrotic changes and kidneys show grayish white spots due to necrosis.

Disease in Humans

The incubation period is usually 10 days with a range of 4-20 days. Leptospirosis in human beings is manifested by fever, respiratory and ocular problems, gastroenteritis, anemia and jaundice.

Diagnosis

1. **Cultural examination:** Diagnosis can be made by isolation and identification of leptospirae from blood during acute illness and from urine after first week of illness. Leptospirae can be isolated by using semisolid medium like Fletcher's medium which consists of nutrient agar and rabbit serum. Solid culture medium can also be used, on which they produce discrete hemispherical colonies just below the surface of the medium in petri dishes.

2. **Serological examination:** Serological tests are commonly used for the diagnosis of leptospirosis. Agglutination test and ELISA are more frequently used serological tests for diagnosis of leptospirosis.

Disease Management in Animals

1. **Vaccination:** A killed mixed vaccine of *L. canicola* and *L. icterohemorrhagiae* is used against canine leptospirosis. Cattle may be vaccinated with a bivalent hardjo/pomona vaccine.

2. **Treatment:** Intramuscular injection of penicillin, streptomycin or tetracycline is recommended for one week.

3. **Hygienic measures:** Strict hygienic measures should be adopted to prevent establishment of infection through urine. Drinking water should be prevented from contamination. Aborted fetus and fetal membranes should be properly disposed.

Disease Management in Humans

1. **Vaccination:** Vaccine is effective if it contains the serovar that predominate in a particular area. Vaccine containing one type of serovar does not protect the infection caused by other serovars.

2. **Chemoprophylaxis:** Doxycycline @ 200mg orally once a week can be given to persons at risk.

3. **Treatment:** Penicillin is the drug of choice but other antibiotics like tetracycline or doxycycline are also effective.

4. **Hygienic measures:** This includes preventing exposure to potentially contaminated water, reducing contamination by rodent control and protection of workers at risks. Control of rodents and disposal of wastes and health education are also important in prevention and control of disease.

5. **Surveillance:** It provides the basis for intervention strategies in human or veterinary public health. Immediate case-based reporting of suspected or confirmed cases from hospital, general practitioner and laboratory to intermediate level is essential. All cases must be investigated since investigation can identify environmental point sources of transmission and lead to control measures. Hospital-based surveillance may give information on severe cases of leptospirosis. Serosurveillance may give information on whether leptospiral infections occur or not in certain areas or populations.

Chapter 16

Listeriosis

Synonyms

Circling disease, Silage disease, Meningoencephalitis

Listeriosis is a serious but rare infection, mainly caused by ingestion of food contaminated with *Listeria monocytogenes*. The disease is known as "circling disease" in young animals. It is considered as an emerging foodborne infection due to changing food habits, technological advancements for longer shelf-life of food products, and the ability of this microorganism to survive and continue to grow even in the refrigerator. It is an infectious disease affecting animals, birds and humans and characterized by encephalitis.

Etiology

Listeria species are found in soil, water, effluents, a large variety of foods, and the faeces of humans and animals. However, only *L. monocytogenes* and, rarely, *L. ivanovii* are pathogenic for humans, out of ten known *Listeria* species. Most of the listerial infections are caused by *L. monocytogenes* serovars 1/2a and 4b. The organism is Gram-positive, rod-shaped and intracellular. The organism is excreted in faeces which enable it to colonize in the soil, plants and surface water. The organism has also been detected in the human faeces. Healthy animals may harbour pathogenic *Listeria* in their gastrointestinal tracts. Vegetables may become contaminated through soil or manure used as fertilizer. Food items which may occasionally contain *L. monocytogenes* include a wide variety of ready-to-eat or raw foods, such as raw milk or meat and their products, raw mushrooms, soft cheese and seafood. The *Listeria* spp. is able to multiply even at high salt concentrations and in acidic conditions. *Listeria* can grow and multiply in some foods at refrigeration temperature.

Epidemiology

The disease occurs worldwide and has been reported in animals, birds and human beings. The incidence of listeriosis varies between 0.1 and 11.3 cases per million per year in different countries. Reports of listeriosis in countries of Southeast Asia are scarce, either because of failure to detect, failure to report, or low incidence rate or failure to consider listeriosis for differential diagnosis by clinicians. The disease remains largely undiagnosed and under-reported. However, *L. monocytogenes* has been found to be one of the etiological factors in causing spontaneous abortions and premature births in India. Asymptomatic carriers play important role in epidemiology of listeriosis. In most animals the disease is sporadic in nature. In humans the disease is found in sporadic and endemic pattern. Silage feeding during winter season is extremely risky. Human infections show a peak incidence during July and August.

Pregnant women, the elderly or individuals with a weakened immune system due to AIDS, leukaemia, cancer, kidney transplant and steroid therapy, are at greatest risk of severe listeriosis. People with AIDS are at least 300 times more likely to get listeriosis than those with a normal immune system. Pregnant women are about 20 times more likely to get listeriosis than other healthy adults. It can result in miscarriage or stillbirth. Newborns may also have low birth weight, septicaemia and meningitis.

Transmission

1. **Ingestion:** It is the most common method of transmission of listeriosis. Animals may get infection through ingestion of food or feed (silage). Human may be infected due to consumption of contaminated raw milk, cheese, meat, poultry, pork and vegetables.

2. **Contact:** Direct contact with diseased animals during parturition may lead to infection.

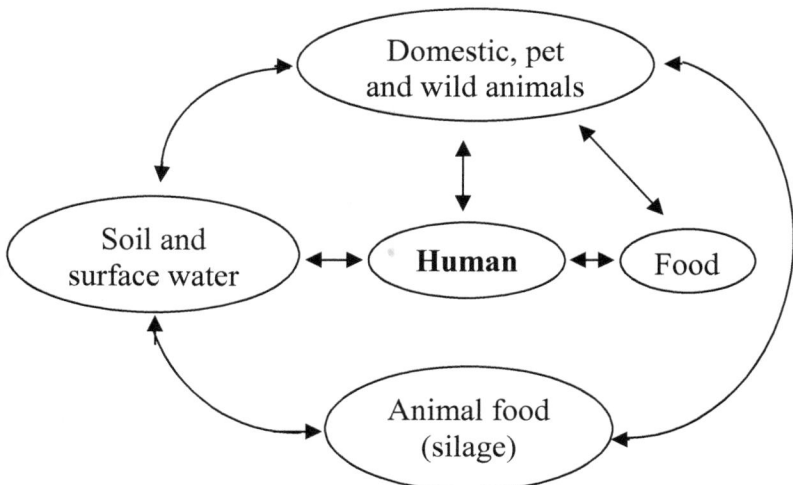

Figure 16.1: Infection Cycle of Listeriosis.

3. **Inhalation:** Airborne infections may occur after inhalation of contaminated dusts from animal houses.

4. **Nosocomial infection:** It has been reported in physicians, midwives, nurses and healthy infants from obstetrical wards through contact.

5. **Trans-uterine transmission:** The disease can also be transmitted to neonates from infected mother during pregnancy.

Disease in Animals

Following ingestion of contaminated food or feed, the organism enters the circulation through intestinal mucosa, where it produces bacteremia. Further the organisms are localized in various organs. The organisms lodge in the brain and produce micro-abscess and hemorrhage. Lesions may extend to meninges and cause meningitis. In animals, the disease is characterized by septicemia, fever, encephalitis, circling, mastitis and abortion. In birds, the disease is characterized by septicemia and necrotic cardiopathy.

Disease in Humans

The incubation period is uncertain but probably ranges from 1 to 4 weeks. The symptoms depend on the physiological and immune status of the patient. A person with mild listeriosis usually has fever and myalgia, preceded by diarrhea or other gastrointestinal symptoms. The most frequent clinical manifestation of severe listeriosis is a neurological syndrome, *e.g.*, severe headache, meningitis or encephalitis. Clinical manifestations in pregnant women include fever, diarrhea, abortion or stillbirth. The organisms affect neonates which mainly cause meningitis. The mortality rate due to infection with *L. monocytogenes* is high (30 per cent). The oculoglandular listeriosis (Parinaud's syndrome) is associated with localized conjunctivitis and swelling of the regional lymph nodes.

Diagnosis

1. **Clinical examination:** Initial diagnosis of listeriosis can be made on the basis of clinical symptoms and presence of the bacteria in a smear from blood, cerebrospinal fluid (CSF), meconium of newborns and vaginal secretions.

2. **Cultural examination:** The organism can be isolated from clinical specimens such as blood, CSF, and the meconium of newborns (or the fetus in abortion cases), as well as faeces, vomits and food stuff/animal feed. During pregnancy, blood cultures and culture of the placenta are the most reliable ways to find out if symptoms are due to listeriosis. For selective isolation of *L. monocytogenes* various plating media such as Polymyxin Acriflavin Lithium chloride Ceftazidime Aesculin Mannitol (PALCAM) agar, Dominguez-Rodriguez agar (DRA) and lithium chloride phenyl ethanol moxalactum (LPM) agar are used. On PALCAM agar, *Listeria* forms black halo around colonies (*i.e.* grey-green in colour with black sunken center and black halo against a cherry-red medium). On DRA, *Listeria* colonies are greenish yellow, glistening pointed, 0.5 mm diameter

surrounded by diffuse black zone of aesculin hydrolysis. On LPM agar, colonies are bluish with ground glass appearance when observed under oblique trans-illumination (Henry's trans-illumination technique).

3. **Serological examination:** Blood tests to detect antibodies are not reliable for diagnosis of listeriosis.

4. **Molecular detection:** Various detection methods including polymerase chain reaction (PCR) are available for diagnosis of listeriosis in humans.

Disease Management in Animals

1. **Treatment:** Treatment with antibiotics like penicillin and tetracycline.

2. **Vaccination:** Killed vaccine can be used to reduce the incidence in sheep. Live vaccine containing serotype 1 and 4B (2 ml subcutaneously at 3 month of age) can be useful in prevention of listeriosis in sheep.

3. **Hygienic measures:** Maintenance of high standards of hygiene at animal farm, regular disinfection of milk parlor and proper management of silage can reduce the incidence of listeriosis.

Disease Management in Humans

1. **Treatment:** Listeriosis is a treatable disease if diagnosed early. When infection occurs during pregnancy, prompt administration of antibiotics can often prevent infection of the foetus or newborn. Depending on the form of the disease, the recommended treatment is amoxycillin with gentamicin. An alternative drug is a combination of trimethroprim and sulfamethoxazole. *Listeria* is naturally resistant to cephalosporins.

2. **Heat treatment of food:** Adequate heat treatment of milk, cheese, poultry, pork and vegetables can reduce the incidence of infection in human beings.

3. **Hygienic measures:** Maintenance of high standards of hygiene during milking of animals, preparation of foods and handling of cases of listeriosis by physicians, nurses, midwives in obstetrical wards may be helpful in reducing the infection. Wash raw vegetables thoroughly with clean water before eating. Wash hands, knives, and cutting boards with soap and water after handling and processing of uncooked foods.

4. **Other measures:** Keep uncooked meats separate from vegetables as well as from cooked foods and ready-to-eat foods. Consume perishable and ready-to-eat foods as soon as possible. Refrigerate or freeze perishable food, prepared food and leftovers within two hours. Keep leftovers for a maximum of 4 days, preferably 2-3 days, and reheat them to an internal temperature of 74°C before consuming them. Refrigerators should be periodically well cleaned, and maintained at 4°C. Read and carefully follow the period of shelflife indicated on the label of the product.

Special Precautions to be taken for Persons at High Risk

1. Avoid milk or dairy products such as soft cheese made from unpasteurized milk.
2. Do not consume luncheon meat, or deli meats, unless they are reheated until steaming hot.
3. Do not consume soft cheeses unless they have labels that clearly state they are made from pasteurized milk.
4. Do not consume refrigerated pates or meat spreads.
5. Do not consume refrigerated smoked seafood.

Chapter 17

Melioidosis

Synonyms

Rodent glanders, Pseudo-glanders, Pseudo-Malleus, Whitmore's disease

Melioidosis is a worldwide disease of animals and humans which is most prevalent in Southeast Asia and northern Australia.

Etiology

The disease is caused by *Burkholderia pseudomallei* (formerly *Pseudomonas pseudomallei*), a saprophytic Gram-negative motile bacillus. The bacteria are found in soil and water, widely distributed in tropical and subtropical countries.

Epidemiology

Melioidosis is endemic in Southeast Asia, Papua, New Guinea, much of the Indian subcontinent, southern China, Hong Kong, and Taiwan. Melioidosis is highly endemic in northeast Thailand, Malaysia, Singapore, and northern Australia. Sporadic cases have been reported among residents of or travelers to Aruba, Colombia, Costa Rica, El Salvador, Guatemala, Guadeloupe, Honduras, Martinique, Mexico, Panama, Venezuela, and many other countries in the Americas, as well as Puerto Rico. In northern Brazil, clusters of melioidosis have been reported and are associated with periods of heavy rainfall.

The risk is highest for adventure travelers, ecotourists, military personnel, construction and resource extraction workers, and other people whose contact with contaminated soil or water may expose them to the bacteria. The infections have been reported in people who have spent less than a week in an endemic area. Risk factors for systemic melioidosis include diabetes, excessive alcohol use, chronic renal disease, chronic lung disease, thalassemia, and malignancy or other non-HIV-related immune suppression.

Transmission

The disease is transmitted through subcutaneous inoculation, ingestion and inhalation. Person-to-person transmission is extremely rare but may occur through contact with the blood or body fluids of an infected person.

Disease in Humans

The incubation period is generally 1-21 days, although it may extend for months or years. Melioidosis may occur as a subclinical infection, localized infection (such as cutaneous abscess), pneumonia, meningoencephalitis, sepsis, or chronic suppurative infection. The latter may mimic tuberculosis, with fever, weight loss, productive cough, and upper lobe infiltrate, with or without cavitation. More than 50 per cent of cases present with pneumonia.

Diagnosis

Culture of *B. pseudomallei* from blood, sputum, pus, urine, synovial fluid, peritoneal fluid, or pericardial fluid is useful for diagnosis. Indirect haemagglutination assay is a widely used serologic test but is not considered confirmatory.

Disease Management in Humans

Travelers should use personal protective equipment such as waterproof boots and gloves to protect against contact with contaminated soil and water and thoroughly clean skin lacerations, abrasions, or burns that have been contaminated with soil or surface water. Ceftazidime, imipenem, or meropenem can be used for initial treatment of 10-14 days, followed by 20–24 weeks of trimethoprim-sulfamethoxazole. Recurrence of disease may occur, especially in patients who received a shorter-than-recommended course of therapy.

Chapter 18

Pasteurelloses

Synonyms

Barbone; In cattle- Hemorrhagic septicemia, Hemorrhagic pneumonia, Stockyard pneumonia, Shipping fever; In pigs- Snuffles; In poultry- Fowl cholera, Pasteurellosis

Pasteurelloses are infectious diseases of animals which may cause disease in humans after contact and biting or scratch by infected animals. Many *Pasteurella* species are opportunistic pathogens that can cause endemic disease and are associated increasingly with epizootic outbreaks.

Etiology

Pasteurelloses are primarily caused by *Pasteurella multocida*. It is also caused by *P. haemolytica*, *P. dagmatis*, *P. stomatis*, *P. canis* and *P. cabali*. The genus *Pasteurella* is a member of the Pasteurellaceae family, which includes a large and diverse group of Gram-negative gammaproteobacteria, whose members are not only human or animal commensal and/or opportunistic pathogens but also outright pathogens. *Pasteurella* spp. is non-motile and rod-shaped.

Epidemiology

Pasteurelloses occur worldwide in domestic and wild animals and birds. *Pasteurella* species are highly prevalent among animal populations, where they are often found as part of the normal microbiota of the oral, nasopharyngeal, and upper respiratory tracts. They cause numerous endemic and epizootic diseases of economic importance in a wide range of domestic and wild animals and birds. *P. multocida* is a common commensal or opportunistic pathogen found in the upper respiratory tracts of most livestock, domestic, and wild animals, including chickens, turkeys, and other wild birds, cattle and bison, swine, rabbits, dogs, cats, tigers, leopards, lions, goats, chimpanzees, marine mammals and komodo dragons. Animal owners, breeders, farmers, veterinarians and abattoir workers are at high risk of acquiring *Pasteurella* infections.

Transmission

The disease transmission in humans usually occurs through animal bites or contact with nasal secretions, with *P. multocida* being the most prevalent isolate observed in human infections. Transmission is through direct contact with nasal secretions, where a chronic infection ensues in the nasal cavity, paranasal sinuses, middle ears, lachrymal and thoracic ducts of the lymph system and lungs. Pre-existing or co-infection with other respiratory pathogens, particularly *Bordetella bronchiseptica* or *Mannheimia haemolytica*, significantly enhances colonization by *P. multocida*, leading to more severe disease.

Disease in Animals

In animals, the disease is characterized by septicemia, pneumonia and atrophic rhinitis. The predominant syndrome of pasteurellosis in endemic and epizootic infections of wild and domestic animal populations is rhinitis, nasal secretions and pneumonia. Symptoms of pasteurellosis in most animals range from mild to severe. Mild symptoms include sneezing, copious mucous secretions, mild rhinitis, mild pneumonia with labored breathing, and fever but can progress to disseminated disease (hemorrhagic septicemia and/or atrophic rhinitis associated with toxinogenic strains. Pasteurellosis pneumonia without symptoms of atrophic rhinitis is most often caused by nontoxinogenic capsular type A strains of *P. multocida*.

P. multocida is often endemic in rabbit colonies and swine herds, where the pneumonia and rhinitis disease is commonly called "snuffles". In more severe cases, symptoms progress toward atrophic rhinitis and, in rare cases, renal impairment, testicular and splenic atrophy, and hepatic necrosis. Atrophic rhinitis in rabbits can also result in overall weight loss, growth retardation, and frequently death. Toxinogenic capsular serotype D and some serotype A strains of *P. multocida* are associated with more severe symptoms of atrophic rhinitis in rabbits and swine. Serotype D is more prevalent in swine and serotype A more prevalent in rabbits. Atrophic rhinitis in swine exhibits mild to severe form caused by infection with toxinogenic *P. multocida*. *P. multocida* subsp. *multocida* is the most predominant cause of fowl cholera worldwide in a variety of avian species, although *P. multocida* subsp. *septica* and *P. multocida* subsp. *gallicida* are also sometimes isolated.

Disease in Humans

The incubation period of pasteurellosis depends on the portal of entry. It may vary from 2 to 14 days. The manifestation and pathological symptoms range from asymptomatic or mild chronic upper respiratory inflammation to acute, often fatal and pneumonic and/or disseminated disease. Common symptoms of pasteurellosis in humans from animal bite wounds are swelling, cellulitis, and bloody or suppurative/purulent exudate at the wound site. In more severe cases, pasteurellosis can rapidly progress to bacteremia and other complications such as osteomyelitis, endocarditis and meningitis.

Respiratory infection in humans is relatively uncommon but can occur in patients with chronic pulmonary disease. This is characterized by severe bilateral consolidating pneumonia, lymphadenopathy, epiglottitis and abscess formation. Neonatal meningitis, usually with septicemia has been reported. Patients with immunocompromised condition have an increased risk of peritonitis, endocarditis, and/or septicemia caused by *P. multocida*.

Diagnosis

1. **Cultural examination:** *Pasteurella* spp. can be isolated from specimens of infected bite wounds, abscess, pus, blood and CSF. *Pasteurella* spp. grows well on sheep or rabbit blood agar in atmosphere of 5 per cent CO2.

2. **Microscopic examination:** *Pasteurella* spp. shows bipolar staining.

3. **Molecular method:** PCR method is useful for diagnosis of disease.

Disease Management in Animals

1. **Avoid stress:** Avoid stress to the animals by providing proper feed, water, shelter, space and transport facilities.

2. **Isolation of animals:** Diseased or newly purchased animals should be isolated from existing and healthy stock.

3. **Hygienic measures:** These include proper disposal of dead animals, prevention of contamination of feed and water and disinfection of articles, feed and premises contaminated with secretions of diseased animals.

4. **Vaccination:** Alum precipitated or aluminium hydroxide gel vaccine can be used.

5. **Use of hyper-immune serum:** It can be used for immediate control of disease during an outbreak.

6. **Treatment:** Sulphadimidine is effective for treatment of this disease. Antibiotics like penicillin, tetracycline or chloramphenicol can also be used as alternative drugs.

Disease Management in Humans

Hygienic measures are very important particularly in preventing contact infections from animals. Penicillin and doxycycline is the drug of choice. Penicillin V @ 500 mg orally, four times a day or doxycycline @ 100 mg orally; twice a day for 10 days can be used for effective treatment of pasteurellosis. Polymicrobial bite wounds can be treated with amoxycillin-clavulanic acid @ 500/100 mg, three times a day.

Chapter 19

Plague

Synonyms

Mahamari, Black death, Mad rat disease

Plague is one of the oldest and most dangerous zoonosis and one of the most virulent and potentially lethal bacterial diseases. It was known as the "Black death" during the fourteenth century, causing an estimated 50 million deaths. The infection is maintained in nature particularly by rodents. Human infection occurs only secondarily. It is transmitted between animals and humans by the bite of infected fleas, direct contact, inhalation and rarely, ingestion of infective materials.

Etiology

The disease is caused by *Yersinia pestis* (formerly *Pasteurella pestis*), which belongs to the family Enterobacteriaceae. *Y. pestis* is considered as potential agent of bioterrorism. It is Gram-negative, non-motile coccobacillus that exhibits bipolar staining with special stain like "Wayson's stain". The bacilli occurs in great abundance in the buboes, blood, spleen, liver and other visceral organs of infected persons and in the sputum of cases of pneumonic plague. It has been observed that plague bacilli can survive and multiply in the soil of rodent burrows where microclimate and other conditions are favourable. Infected rodents and fleas and cases of pneumonic plague are the main source of infection.

Epidemiology

Outbreaks of plague are usually seasonal in nature. In northern India the "plague season" starts from September until May. A mean temperature of 20-25°C and a relative humidity of 60 per cent and above are considered favourable for the spread of plague. Heavy rainfall especially in the flat fields tends to flood the rat burrows. This factor may be responsible for keeping certain states like West Bengal free from plague. Rats frequent dwelling houses and where housing conditions are poor, there may be an abundance of rats and rat fleas all the around year and contact with human occurs readily.

Plague occurs in many forms such as epizootic, enzootic, sporadic and in epidemic of all types including anthroponotic and primary pneumonic forms. All age groups and both sexes of human beings are susceptible for plague. Plague is endemic in many countries in Africa, the former Soviet Union, the Americas and Asia. The 3 most endemic countries are Madagascar, the Democratic Republic of Congo and Peru. Plague epidemics have occurred in Africa, Asia, and South America but since the 1990s, most human cases have occurred in Africa.

The distribution of plague coincides with the geographical distribution of the rodents. The human may come into contact with natural foci of plague during the course of hunting, cultivation, harvesting or outdoor recreation activities. The natural foci of plague exist in India, Indonesia, Myanmar and Nepal. In India the annual mortality was over 5, 00,000 deaths between 1898 and 1908. The disease continued to be major public health problem till 1940. Uttar Pradesh became free of plague in 1957 and Madhya Pradesh in 1960. Reappearance of plague after a gap of long period that is return of plague has also been reported. Since the last reported cases in 1966, there have been no laboratory confirmed cases of human plague in India, till its appearance in September, 1994 in which 4 persons were found positive for bubonic plague in Beed (Maharashtra) followed by an outbreak of pneumonic plague in Surat (Gujarat). Cases were also reported from Delhi, Mumbai, Kolkata and some other places. Among 4,780 suspected plague cases, 167 cases were found positive with 53 deaths. In an outbreak of plague in February, 2002 in Hat Koti village, Shimla, Himachal Pradesh, the Ministry of Health reported a total of 16 cases of pneumonic plague including 4 deaths. Total 11479 cases of human plague and 772 deaths were reported by 14 countries in Africa, the Americas and Asia from 2004 to 2008.

Flea Indices

Flea indices are useful measurements of the density of fleas. Flea indices do not themselves indicate an imminent plague epidemic. They serve as a warning that more stringent control measures are needed to protect the human population. The following indices are widely used in rat flea surveys:

1. **Total flea index:** It is the average number of fleas of all species per rat.

2. **Cheopis index:** It is the average number of *Xenopsylla cheopis* per rat. If cheopis index is more than one, it is regarded as explosive situation regarding outbreak of plague.

3. **Burrow index:** It is the average number of free-living fleas per species per rodent burrow.

4. **Specific percentage of fleas:** It is the percentage of different species of fleas that are found on rats.

Transmission

Wild rodents such as field mice, gerbils and skunks are the natural reservoir of plague. In India, the wild rodents *Tatera indica* has been incriminated as the main reservoir, not the domestic rat *Rattus rattus* as once thought. The commonest and most efficient vector of plague is the rat flea the *Xenopsylla cheopis* but other fleas such as *X. astia*, *X. braziliense* and human fleas like *Pulex irritans* can also transmit the infection. Both sexes of fleas bite and transmit the infection. A flea may ingest up to 0.5 mm cubic blood which may contain about 5,000 plague bacilli. The bacilli multiply enormously in the gut of the rat flea and may block the proventiculus so that no food can pass through. Such a flea is called "blocked flea". The blocked flea eventually faces starvation and death because it is unable to obtain a blood meal. It makes frantic effort to bite and suck blood over and over again and inoculate (regurgitate) plague bacilli into the bite wound each time of bite. A partially blocked flea is more dangerous than a completely blocked flea because it can live longer. Human plague is most frequently contracted in the following ways-

1. By infected flea bite.
2. By droplet infection from cases of pneumonic plague.
3. Occasionally by direct contact with the tissues of infected animal.

Note: Transmission of plague from plague patient by the bite of infected human flea is a rare and exceptional occurrence.

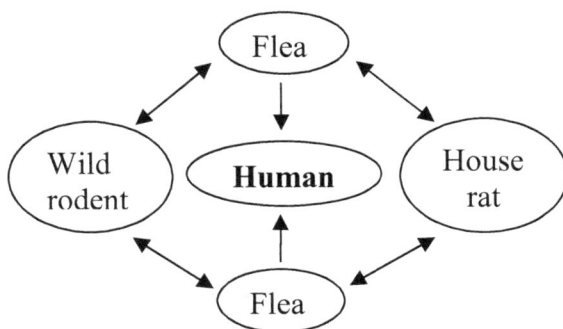

Figure 19.1: Infection Cycle of Plague.

Incubation Period

It varies in different forms of plague as follows:

1. Pneumonic plague: 1- 3 days.
2. Bubonic plague: 2-7 days.
3. Septicemic plague: 2-7 days.

Silent Period

Silences of long duration (10 years or more) followed by sudden explosive outbreaks of rodent or human plague have been repeatedly confirmed in some natural foci (by infected rodents, fleas or migrating birds). It has been observed that the plague disappears completely for long periods and reintroduces from other areas.

Disease in Animals

Plague is primarily a disease of rodents in which human becomes accidentally involved. Animal disease is similar to that in human. Plague is epizootic and enzootic in wild rodents. The two ecological cycles of plague have been described-

1. **Wild plague:** It is defined as "plague existing in nature, independent of human population and their activities". The disease is transmitted among wild rodents by wild rodent fleas. The epizootic wipes out the susceptible population. Those that survive that are resistant species maintain the enzootic in natural foci.

2. **Domestic plague:** It is closely associated with human and rodents living with him, and has a definite potential for producing epidemics.

Disease in Humans

People infected with plague usually develop "flu-like" symptoms after an incubation period of 3-7 days. Typical symptoms are the sudden onset of fever, chills, head and body-aches, weakness, nausea and vomiting. Plague can be a very severe disease in people, with a case-fatality rate of 30 per cent -60 per cent, if left untreated. There are 3 forms of plague infection depending on the route of infection: bubonic, pneumonic and septicemic.

1. **Bubonic plague:** Bubonic plague (known in mediaeval Europe as the 'Black death') is the most common form of plague and is caused by the bite of an infected flea. The infected rat fleas usually bite on the lower extremities and inoculate the plague bacilli. Plague bacilli enter at the bite and travel through the lymphatic system to the nearest lymph node where they are intercepted by the regional lymphatic glands and multiply. Usually within a few days inflamed, tense, painful and greatly enlarged tender lymph nodes (buboes) develop in the groin region and less often in the axila or neck, depending on the site of the bite by the flea. At advanced stages of the infection, the inflamed lymph nodes can turn into suppurating open sores. Typically the patient develops sudden onset of fever, chills, headache, prostration and painful adenitis. Bubonic plague cannot spread from person to person because plague bacilli are locked in the buboes and do not find easy way to exit.

2. **Pneumonic plague:** It is the most virulent and least common form of plague. Primary pneumonic plague is rare. The incidence of pneumonic plague is usually less than 1 per cent. Typically, the pneumonic form is caused by spread to the lungs from advanced bubonic plague. However, a person with secondary pneumonic plague may form aerosolized infective droplets and transmit plague via droplets to other humans. Untreated pneumonic plague has a very high case fatality rate.

3. **Septicemic plague:** Primary septicemic plague is rare except for accidental laboratory infections. However bubonic plague may develop into septicemic plague. Septicemic plague occurs when infection spreads directly through the bloodstream without forming a "bubo". Septicemic plague may result from flea bites and from direct contact with infective materials through cracks in the skin. Advanced stages of the bubonic form of plague will also lead to direct spread of *Y. pestis* in the blood.

Diagnosis

Medical practitioners should keep in mind in differential diagnosis of any cases of fever with lymphadenopathy or when multiple cases of pneumonia occur. There are several methods for diagnosis of plague but confirmatory diagnosis can be made by isolation and identification of plague bacilli from rodents or humans.

1. **Cultural examination:** The best way to confirm that a patient has plague is to identify *Y. pestis* in a sample of fluid from a bubo, or blood or sputum. Colonies on blood agar are at first very small, transparent, white circular discs, 1 mm or less in diameter but later enlarge to 3-4 mm and become opaque.

2. **Morphological examination:** Smear examination from clinical materials such as bubo fluid, sputum *etc.* stained with "Giemsa stain" or "Wayson's stain" can be done to demonstrate bipolar staining of bacilli in the specimen.

3. **Serological examination:** Serological examination of blood sera from acute and convalescent patients can be done for antibodies study. Immunofluorescent microscopic test is important.

4. **Animal inoculation test:** It can be performed using guinea pigs or mice.

5. **Other methods:** Rapid dipstick tests have been validated for field use to quickly screen for *Y. pestis* antigen in patients.

Diseases Management in Humans

The disease can be managed by adopting the following measures:

1. **Control of cases:** Cases can be controlled in the following ways:
   - ☆ **Early diagnosis-** It is important to take necessary action for preventing the complications of disease.
   - ☆ **Notification-** It is important to prevent the rapid spread of the disease. If a human or rodent case is diagnosed, the health authorities must be notified promptly. Case notification is required by International Health Regulations.
   - ☆ **Isolation of patients-** It is important to isolate the patients with pneumonic plague in order to prevent the spread of infection in other persons.

2. **Control of fleas:** Insecticides such as DDT and BHC can be used as dusts containing 10 per cent and 3 per cent of the active ingredients, respectively. In areas where resistance to one or both insecticides occur, dusts of 2 per cent carbaryl or 5 per cent malathion should prove effective. Rat burrows should be insufflated with insecticidal dust with the help of a dust blower.

3. **Control of rodents:** Rodents can be controlled by different measures such as sanitation, fumigation, trapping, and use of chemosterilants and rodenticides.

4. **Identification and control of source of infection:** Identify the most likely source of infection in the area where the human case(s) was exposed, typically looking for clustered areas with large numbers of small animal deaths. Institute the appropriate sanitation and control measures to stop the exposure source.

5. **Disinfection:** Disinfection of sputum, discharges and articles soiled by the patient should be carried out. Dead bodies should be handled with aseptic precautions.

6. **Health education:** Health education to the public about prompt reporting and of dead rats and suspected human cases is essential in order to take preventive measures. Preventive measures include informing people when zoonotic plague is active in their environment and advising them to take precautions against flea bites and not to handle animal carcases in plague-endemic areas. People should also avoid direct contact with infected tissues like suppurating buboes or exposure to patients with pneumonic plague.

7. **Vaccination:** Plague vaccines were once widely used but have not been shown to be very effective against plague. Vaccines are currently not recommended during outbreaks but are still used for high-risk groups (*e.g.,* laboratory personnel who are constantly exposed to the risk of contamination). Vaccination should be carried out at least one week before an anticipated outbreak of plague. Two doses of 0.5 ml and 1.0 ml are administered at interval of 1-2 weeks. Booster doses are administered six monthly in persons at continuous risks of infection. Immunity starts one week after inoculation and lasts for about six months.

8. **Chemoprophylaxis:** It is recommended to persons at continuous risks of infection. The drug of choice is tetracycline (500 mg, 6 hourly for 5 days).

9. **Treatment:** Untreated plague can be rapidly fatal so early diagnosis and treatment is essential for survival and reduction of complications. Antibiotics and supportive therapy are effective against plague if patients are diagnosed in time. Treatment of the patient must be started without waiting confirmation of the diagnosis. Unless promptly treated, plague may have a mortality of nearly 50 per cent and pneumonic plague 100 per cent. The drug of choice is streptomycin (30 mg/kg body weight, daily, i.m., in two divided doses for 7-10 days. Tetracycline or sulphonamide can be used as alternative drugs.

10. **Surveillance:** Surveillance and control requires investigating animal and flea species implicated in the plague cycle in the region and developing environmental management programmes to limit spread. Active long-term surveillance of animal foci, coupled with a rapid response during animal outbreaks has successfully reduced numbers of human plague outbreaks. Surveillance is essential in areas where natural foci of plague exists or areas of known history of past infection.

Chapter 20

Rat Bite Fever (RBF)

Synonyms

Sodoku, Haverhill fever, Spirillosis, Streptobacillosis

Rat-bite fever is an infectious disease that can be caused by two different bacteria. In Japan, the disease is known as Sodoku. People usually get the disease from infected rodents or consumption of contaminated food or water. When the latter occurs, the disease is often known as Haverhill fever. Untreated RBF can be a serious or even fatal disease.

Etiology

Streptobacillary RBF is caused by *Streptobacillus moniliformis* in North America, while spirillary RBF is caused by *Spirillum minus* and occurs mostly in Asia. *Streptobacillus moniliformis* is a fastidious, non-motile, Gram-negative microaerophilic bacillus. *Spirillum minus* is a short, thick and motile spirochete.

Epidemiology

The disease occurs mainly in North America and Asia. The disease is rare in the United States. Persons having contact with rodents at home or in the workplace are at potential risk for RBF. Some people who may be at increased risk include those who live in rat-infested buildings, have pet rats at their home and work with rats in laboratories or pet stores.

Transmission

S. moniliformis and *S. minus* are part of the normal respiratory flora of rodents. Organism may be transmitted to humans through bites or scratches. Infection can also result from handling an infected rodent (even with no reported bite or scratch), or ingestion of food or drink contaminated with these bacteria (Haverhill fever).

Rats are considered the natural reservoir of RBF agents, but the bacterium has also been found in other rodent species such as mice and gerbils. Person-to-person transmission has not been reported.

Disease in Humans

Symptoms are often different for the two types of RBF that is Streptobacillary RBF and Spirillary RBF. The initial symptoms of *S. moniliformis* are non-specific and include fever, chills, myalgia, headache and vomiting. There may be development of maculopapular rash on the extremities after 2-4 days of onset of fever, followed by polyarthritis in approximately 50 per cent of patients. The incubation period typically ranges from 3-10 days. The symptoms of Haverhill fever differ slightly from those of RBF acquired through bites and/or scratches. Haverhill fever can be associated with more severe nausea/vomiting and pharyngitis. Symptoms due to *S. minus* usually occur 7-21 days after exposure to an infected animal and the patient is likely to have a history of international travel. Following partial healing of the rat bite, common symptoms include fever, ulceration at the site, lymphangitis, lymphadenopathy and a distinct rash of purple or red plaques. Untreated infection may result in endocarditis, myocarditis, meningitis, pneumonia or sepsis. The mortality rate may be 7 per cent -13 per cent in untreated RBF cases.

Diagnosis

1. **Clinical examination:** RBF should be suspected in people with rash, fever, and arthritis and a known or suspected history of rodent exposure.

2. **Cultural examination:** *S. moniliformis* is difficult to grow in culture and requires specific media and incubation conditions. Organism can be isolated from blood, synovial fluid, or other body fluids. In the absence of a positive culture, identification of pleomorphic Gram-negative bacilli in appropriate specimens supports a preliminary diagnosis.

3. **Microscopic examination:** *S. minus* does not grow in artificial media. For this reason, diagnosis is made by identifying characteristic spirochetes in appropriate specimens using darkfield microscopy or differential stains.

Disease Management in Humans

1. Avoid the contact with rodents and their habitats.
2. Avoid the consumption of milk or water that may have come in contact with rodents.
3. Consume the pasteurized milk and water from safe sources.
4. Wear the protective gloves while handling rats.
5. Practice the regular hand washing while handling the rats or clean their cages.

Chapter 21

Rickettsioses

Rickettsioses are group of infections caused by rickettsiae and transmitted by arthropod vectors. Rickettsiae are obligate, intracellular, Gram-negative coccobacilli and are found within the cytoplasm and occasionally the nucleus of eukaryotic cells. Rickettsial zoonoses may occur as sporadic or endemic diseases. Rickettsial diseases are grouped as spotted fever group, typhus group and scrub typhus group. Some important rickettsial zoonoses are discussed as follows:

Rocky Mountain Spotted Fever (RMSF)

Rocky Mountain spotted fever is an important rickettsiosis in the Western hemisphere. It is a tick-borne rickettsial disease mainly in North and South America. The disease is characterized by fever, headache, rash, vomiting, abdominal pain and myalgia.

Etiology

The disease is caused by *Rickettsia rickettsii.*

Epidemiology

The disease occurs in Western hemisphere, mainly in North and South America. The important reservoirs are rodents, dogs and foxes. The disease is endemic in US, Canada, Mexico, Costa Rica, Columbia and Brazil. The disease does not exist in Europe, Asia and Africa.

Transmission

RMSF is transmitted by infected tick bites. The important tick vectors are *Dermacentor andersonii, D. variabilis, Amblyomma cayennense* and *Rhipicephalus sanguineous.*

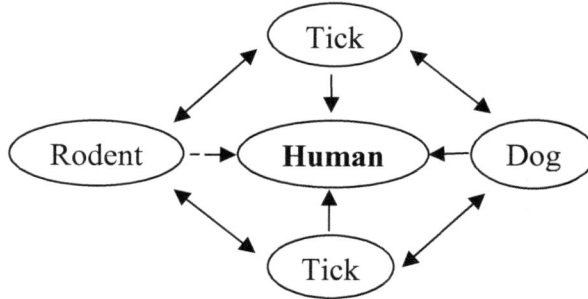

Figure 21.1: Infection Cycle of RMSF.

Disease in Humans

The incubation period of the organism may vary from 3 to 12 days. The disease is characterized by sudden onset of fever, headache, malaise, myalgia, anorexia, nausea, vomiting and abdominal pain. Rash develops in most of the cases, but often absent in first few days of illness. Approximately 10 per cent of RMSF patients never develop a rash. The rash develops which is initially pink. Later, it becomes maculopapular, petechial and hemorrhagic. A classic case of RMSF involves a rash that first appears 2-5 days after the onset of fever as small, flat, pink, non-itchy spots on the wrists, forearms, and ankles and spreads to include the trunk and sometimes the palms and soles. The rash becomes more intensive when body temperature rises. The complication of disease may lead to shock, pulmonary edema, gangrene, acute respiratory dress syndrome, hepatosplenomegaly, hemorrhages in intestinal and urogenital tracts, coma and convulsion.

Diagnosis

1. **Clinical examination:** The disease can be diagnosed on the basis of clinical symptoms. The disease may be suspected even in the absence of rash in initial few days.

2. **Cultural examination:** The disease can be diagnosed by culture of the organism from plasma or buffy coats in Vero, L929 or MRC5 cells.

3. **Serological examination:** In 85 per cent patients, the antibodies are not detectable in the first week of illness, but a negative test during this period does not rule out RMSF as a cause of illness. The gold standard serologic test for diagnosis of RMSF is the indirect immunofluorescence assay (IFA) with *R. rickettsii* antigen. IFA is performed on two paired serum samples to demonstrate a significant (four-fold) rise in antibody titers. The first sample should be taken as early in the disease as possible, preferably in the first week of symptoms, and the second sample should be taken 2-4 weeks later.

4. **Molecular examination:** The organism infects the endothelial cells that line blood vessels, and does not circulate in large numbers in the blood unless the patient has progressed to a very severe phase of infection. For

this reason, blood specimens (whole blood, serum) are not always useful for detection of the organism through polymerase chain reaction (PCR) or culture. If the patient has a rash, PCR or immunohistochemical (IHC) staining can be performed on a skin biopsy taken from the rash site.

Disease Management in Humans

1. **Control of ticks:** Ticks can be controlled by using acaricides.
2. **Protective measures:** Use of protective clothing and tick repellents are important. Use of tick collars on dogs is also important. Avoid exposure to natural foci.
3. **Treatment:** The disease can be fatal if not treated in the first few days of illness. The drug of choice is doxycycline (@ 100 mg, twice a day, orally for 1-2 weeks).

Mediterranean Spotted Fever (MSF)

Synonym

Boutonneuse fever

Mediterranean spotted fever (MSF) is an important rickettsial disease, caused by *Rickettsia conorii* and transmitted by the dog tick *Rhipicephalus sanguineus*. The tick bite causes a characteristic rash and a distinct mark, namely, a tache noire (black spot) at the site of the bite.

Etiology

The disease is caused by *Rickettsia conorii*. It is an obligate, intracellular, Gram-negative bacterium.

Epidemiology

MSF was first reported in Tunisia by Conor and Bruch and was soon reported in other regions around the Mediterranean basin. The disease was thereafter also known as Boutonneuse fever (spotted fever) because of the manifestation of a papular rather than a macular rash. The organism has been reported from southern Europe, Morocco, Israel, Georgia, Ethiopia, Kenya, South Africa, India and Pakistan. MSF is endemic in Mediterranean area, including northern Africa and southern Europe. MSF is an emerging or a re-emerging disease in some countries. The first case of MSF was clinically diagnosed in 1993in Oran, Algeria. Since that time, the number of cases has steadily increased. The incidence of disease has increased in the past 10 years in some other countries of the Mediterranean basin such as Italy and Portugal. The important reservoirs of MSF agent are rodents and dogs.

Transmission

The disease is transmitted by infected tick bites. The important tick vector of the disease is *Rhipicephalus sanguineous*.

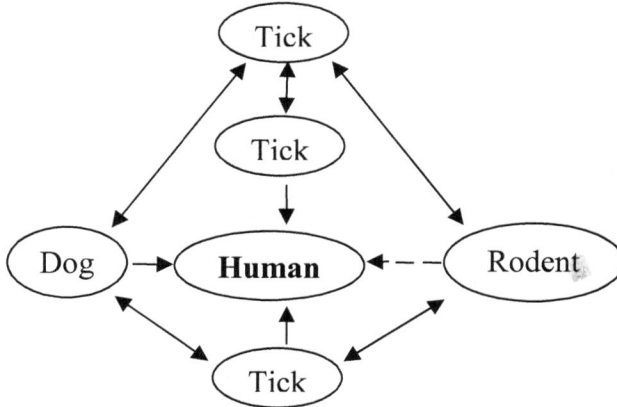

Figure 21.2: Infection Cycle of MSF.

Disease in Humans

The incubation period of the organism may vary from 2 to 7 days. The disease is characterized by development of primary lesion with central necrosis at the site of tick bite. The lesion is covered with a brownish black scab which may ulcerate. In some patients, the eruption is papulovesicular; this form is more common in adults in Africa. In other patients, the only symptom is an isolated lymphadenopathy. Other symptoms are fever, headache, arthralgia, myalgia, regional lymphadenitis and generalized maculopapular rash. The complication of the disease may lead to meningoencephalitis, coma and heart failure.

Diagnosis

1. **Clinical examination:** The disease can be diagnosed on the basis of history of tick bite and clinical symptoms like tache noire and rash.

2. **Immunohistochemical examination:** The organism can be detected in skin biopsy specimens from area of rash.

3. **Serological examination:** Serological tests like complement fixation test, Immunofluorescence assay and enzyme immunoassay are useful for diagnosis of the disease.

4. **Molecular examination:** PCR can be used for detection of the organism from peripheral blood or tissue specimens.

5. **Isolation of the organism:** Rickettsiae can be isolated from peripheral blood, buffy coat or tissue specimens.

6. **Animal inoculation test:** Intraperitoneal inoculation of infected material in the guinea pig induces fever, scrotal edema and serofibrinous exudate in periorchium.

Disease Management

1. **Protective measures:** Avoid exposure to natural foci of the disease.

Protective clothing can also be used. Use of tick collars on dogs can be very effective prophylactic measure.

2. **Control of ticks:** Ticks can be controlled by using acaricides.

3. **Treatment:** The drug of choice is doxycycline (@ 100 mg orally, twice daily for one week). Alternative antibiotics are tetracycline and chloramphenicol.

Epidemic Typhus

Synonyms

Louse-borne typhus, European typhus, Classic typhus, Brill-Zinsser disease, Jail fever

Epidemic typhus is a highly fatal rickettsial disease. The disease can cause explosive epidemics in humans. In the past, the disease was associated with wars and human disasters and it is still endemic in the highlands and cold areas of Africa, Asia and Central and South America. The disease is characterized by the sudden onset of high fever, headache, chills, prostration, coughing and severe myalgia. A macular eruption appears on the fifth to sixth day.

Etiology

The disease is caused by *Rickettsia prowazekii*. The organism is antigenically related to *R. typhi* and *R. canada*.

Epidemiology

Epidemic typhus occurs in colder regions of Central and Eastern Africa, Central and South America, and Asia. In recent years, most outbreaks have been reported from Burundi, Ethiopia and Rwanda. Humans are the only reservoir and are responsible for maintaining the infection during inter-epidemic periods. Infections have also been reported in squirrels and lice harbouring on them. The disease occurs in conditions of overcrowding and poor hygiene, such as in prisons and refugee camps. Humanitarian relief workers may be exposed in refugee camps and other settings characterized by crowding and poor hygiene. The body louse lives in clothing and multiplies very rapidly under poor hygienic conditions. Since World War II, large outbreaks of typhus have occurred mainly in Africa, with reported cases coming predominantly from Burundi, Ethiopia and Rwanda. Sporadic cases have also been reported from US. Louse-borne typhus should be suspected when people in crowded, louse-infected conditions manifest sudden onset of high fever, chills, headache, bodyache, exhaustion, agitation, followed on the fifth or sixth day by a macular eruption.

Transmission

The disease is transmitted by human body louse (*Pediculus humanus corporis*). It is not directly transmitted from person to person. Patients are infective for lice during the febrile illness and possibly for 2-3 days after the temperature returns to normal. Infected lice pass rickettsiae in their faeces within 2-6 days after the blood meal. The louse invariably dies within two weeks after infection but rickettsiae may remain viable in the dead louse for weeks. Sporadic cases may occur due to transmission of infections in squirrels and lice/fleas harbouring on them.

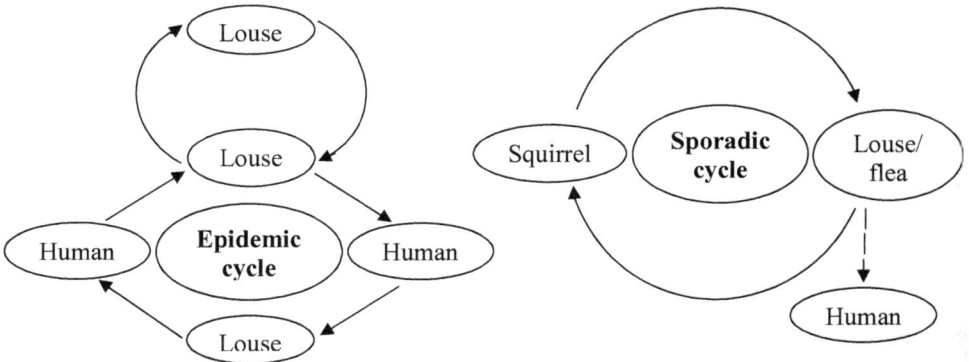

Figure 21.3: Infection Cycle of Epidemic Typhus.

Disease in Humans

Milder symptoms of louse-borne typhus can occur years after the primary attack (Brill-Zinsser disease). The case-fatality rate is between 1 per cent and 20 per cent.

Diagnosis

1. **Cultural examination:** The organism can be cultured in cell culture.

2. **Serological examination:** Clinical diagnosis may be confirmed by serology. Immunofluorescence assay and ELISA are important serological tests for diagnosis of the disease. A fourfold or greater increase in titre or the presence of IgM antibody confirms the diagnosis.

3. **Molecular examination:** The disease can also be diagnosed by using PCR technique.

Disease Management in Humans

1. **Hygienic measures:** The disease can be prevented by adopting high standards of hygiene.

2. **Control of louse:** Suitable insecticidal dusts for body louse control are permethrin (1 per cent), propoxur (1 per cent), temephos (2 per cent) and carbaryl (5 per cent). The WHO recommends the use of 1 per cent permethrin powder for treatment of clothing, in which clothing should be aerosolized in a compressor by using permethrin powder (@ 125 to 250 mg per m² of clothing.

3. **Vaccination:** It is recommended for persons at greater risk of acquiring infection. Live attenuated and a formaldehyde-inactivated vaccines containing *Rickettsia prowazekii* are available.

4. **Chemoprophylaxis:** Doxycycline can be used @ 100 mg, orally, per week in persons in endemic areas.

5. **Treatment:** The drug of choice is doxycycline (@ 100 mg, orally, twice daily until 2 or 3 days after defervescence). Tetracycline and chloramphenicol can be used as alternative drugs.

Murine Typhus

Synonyms

Endemic typhus, Toulon typhus, Rat flea typhus, Urban typhus of Malaya, Tabardillo

Murine typhus is a rickettsial disease of rodents which can be transmitted to humans. The disease is characterized by fever, chills, headache and rash. The disease is clinically similar to but milder than epidemic typhus.

Etiology

The disease is mainly caused by *Rickettsia typhi* (formerly called *Rickettsia mooseri*). The disease can also be caused by *Rickettsia felis*.

Epidemiology

The disease primarily occurs in tropics and subtropics in a sporadic pattern or as miniepidemics. The disease has been reported from Malta, Greece, Yugoslavia, California and Texas. Animal reservoirs include wild rats, mice, and other rodents. Rat fleas and probably cat fleas transmit organisms to humans through bites. Fleas are also natural reservoirs for *R. typhi* and infected female fleas can transmit organisms to their progeny. The peak season of the disease is summer and fall. The incidence of disease is higher in rat-infested areas. The prevalence of the disease is higher in men than women.

Transmission

Murine typhus is transmitted to humans by infected fleas bite. The important vectors are rat fleas like *Xenopsylla cheopis* and *Leptopsylla segnis*. Cat flea (*Ctenocephalides felis*) can also transmit the disease.

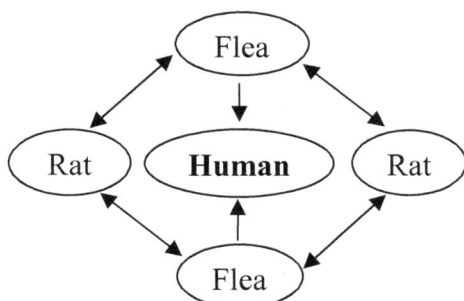

Figure 21.4: Infection Cycle of Murine Typhus.

Disease in Humans

The incubation period is 6 to 18 days (average 10 days). The disease is characterized by onset of shaking chill, fever and headache. The fever lasts about 12 days; then temperature gradually returns to normal. The rash and other manifestations are similar to those of epidemic typhus but are much less severe. The early rash is sparse and discrete. The mortality rate is low (1-8 per cent).

Diagnosis

1. **Clinical examination:** The disease can be diagnosed by clinical symptoms but differentiation from epidemic typhus is difficult.
2. **Immunohistochemical examination:** The organism can be detected in skin biopsy specimens from area of rash by using fluorescent antibody staining. Weil-Felix reaction with *Proteus vulgaris* OX-2 and OX-19 is positive from the 10th day on.
3. **Molecular examination:** The organism can be detected by using PCR technique.

Disease Management in Humans

1. **Control of reservoirs:** Rodents are important reservoirs; therefore, they should be controlled by using rodenticides, mouse traps *etc.* for prevention of spread of the disease.
2. **Control of vectors:** Rat fleas as well as cat fleas can be controlled by using insecticides.
3. **Treatment:** The drug of choice is doxycycline (100 mg, orally, twice daily). The treatment is continued until 2 or 3 days after defervescence. Tetracycline and chloramphenicol can also be used as alternative drugs.

Scrub Typhus

Synonyms

Tropical typhus, Mite-borne typhus, Tsutsugamushi fever, Chigger fever

Scrub typhus is a rickettsial disease particularly in Asia-Pacific. The disease is transmitted by the bite of infected mite larvae (chiggers) and characterized by fever, chills, severe headache, generalized lymphadenopathy and a rash. The term "scrub" is used because of the type of vegetation (terrain between woods and clearings) that harbours the vector. However, the name can be misleading, as endemic areas can also be sandy, semi-arid and mountain deserts.

Etiology

The disease is caused by *Orientia tsutsugamushi*, which is a Gram-negative bacterium of the family Rickettsiaceae.

Epidemiology

Scrub typhus is endemic in an area of Asia-Pacific bounded by Japan, Korea, China, India, and northern Australia. The important reservoirs are rats, mice, rabbits and marsupials. The important vectors are trombiculid mites (*Trombicula* spp.). The natural habitat of the vectors is the areas of moist climates with underbrush.

Transmission

Orientia tsutsugamushi is transmitted by trombiculid mite larvae (chiggers), which feed on forest and rural rodents, including rats, voles, and field mice. Human infection also follows a chigger bite. The mites are both the vector and the natural reservoir for *O. tsutsugamushi*.

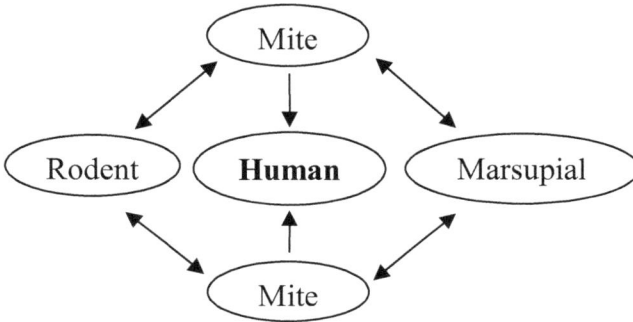

Figure 21.5: Infection Cycle of Scrub Typhus.

Disease in Humans

The incubation period of the organism varies from 6 to 21 days (average 10 to 12 days). The disease is characterized by sudden onset of fever, chills, headache, and generalized lymphadenopathy. At onset of fever, an eschar often develops at the site of the chigger bite. The typical lesion of scrub typhus, common in whites but rare in Asians, begins as a red, indurated lesion about 1 cm in diameter. The lesion eventually vesiculates, ruptures, and becomes covered with a black scab. A macular rash develops on the trunk during the 5th to 8th day of fever, often extending to the arms and legs. It may disappear rapidly or become maculopapular and intensely colored. Cough is present during the first week of fever, and pneumonitis may develop during the second week. In severe cases, pulse rate increases; BP drops; and delirium, stupor, and muscular twitching develop. Splenomegaly may be present, and interstitial myocarditis is more common than in other rickettsial diseases.

Diagnosis

1. **Clinical examination:** The disease can be suspected from history of flea bite and clinical symptoms.

2. **Isolation of the organism:** The organism can be isolated from blood by intraperitoneal injection into mice.

3. **Serological examination:** Biopsy of rash with fluorescent antibody staining is used to detect the organisms. Indirect immunofluorescence assay, passive haemagglutination test and ELISA can also be used. Weil-Felix reaction with *Proteus vulgaris* OX-K is positive in about 50 per cent cases.

4. **Molecular examination:** A PCR has been developed for detection of the organism.

Disease Management in Humans

1. **Control of reservoirs:** Rodents can be controlled by using rodenticides or mouse traps.

2. **Control of vectors:** Clearing bushes and spraying infested areas with residual insecticides eliminate or decrease mite populations. Insect repellents like diethyltoluamide (DEET) should be used when exposure is likely.

3. **Treatment:** Doxycycline (@100 mg, orally, twice daily), tetracycline (@500 mg, orally, four times a day) or chloramphenicol (@500 mg, orally, four times a day) for at least one week is effective.

Q Fever

Synonyms

Rickettsiosis, Burnet's rickettsiosis, Coxiellosis, Query fever, Abattoir fever, Balkan fever, Balkan grippe, Nine-mile fever

Q fever (Q is for query) is a highly infectious zoonotic disease. It occurs mainly in persons associated with cattle sheep, goat or other domestic animals. It may occur as occupational direct anthropozoonosis or metazoonosis. Q fever is a worldwide disease with acute and chronic stages. Cattle, sheep, and goats are the primary reservoirs, although a variety of species may be infected. *C. burnetii* is a highly infectious agent that is rather resistant to heat and drying. A single *C. burnetii* organism may cause disease in a susceptible person. This agent has a past history of being developed for use in biological warfare and is considered a potential terrorist threat.

Etiology

The disease is caused by *Coxiella burnetii*. It is a Gram-negative, short rod and intracellular organism. Domestic animals harbor the *Coxiella* organisms without showing clinical manifestations. The infected animals shed the organisms in the milk, faeces, urine which heavily contaminate the soil. *C. burnetii* is a potential agent of bioterrorism. During parturition the organisms are shed in high numbers within the amniotic fluids and the placenta. The organism is extremely hardy and resistant to heat, drying, and many common disinfectants which enable the organism to survive for long periods in the environment.

Epidemiology

Q fever was first recognized as a human disease in Australia in 1935 and in the United States in the early 1940's. The "Q" stands for "query" and was applied at a time when the causative agent was unknown. Human Q fever is now known to be the result of infection with the obligate, intracellular bacterium, *C. burnetii*. Cattle, sheep, and goats are commonly infected and may transmit infection to humans when they give birth. *C. burnetii* can survive for long periods of time in the environment, and may be spread by wind and dust. The disease occurs as sporadically or occasionally as outbreaks. The important reservoirs of this organism are cattle, sheep, goat, camel, horse, dog, some wild animals and ticks.

The disease occurs worldwide with the exception of New Zealand. The hosts of the organism are ticks, rodents, birds, most of the domestic animals such as cattle, sheep, goat and dog and human beings. The natural reservoirs of *C. burnetii* are ticks that transmit the organism transovarially. The ticks or their excreta spread the disease to domestic animals. The placenta and amniotic fluid of infected animals contain large number of organisms that contaminate the pastures and soil. The serological surveys indicate that the disease is present in the human and animal population in Punjab, Haryana, Rajasthan, Delhi and many other states of India. It is an occupational zoonotic disease. It occurs in the farmers, shepherds, dairy and abattoir workers, laboratory personnel and veterinarians.

Transmission

1. **Inhalation:** The disease in human beings may be transmitted through inhalation of dust contaminated with faeces or urine of infected animals. The disease may also be transmitted through contaminated aerosol.

2. **Ingestion:** The disease in humans may be transmitted through ingestion of contaminated meat, milk and milk products. The disease in domestic animals is transmitted through grazing of pastures contaminated with excreta of infected ticks or placenta and amniotic fluid of infected animals.

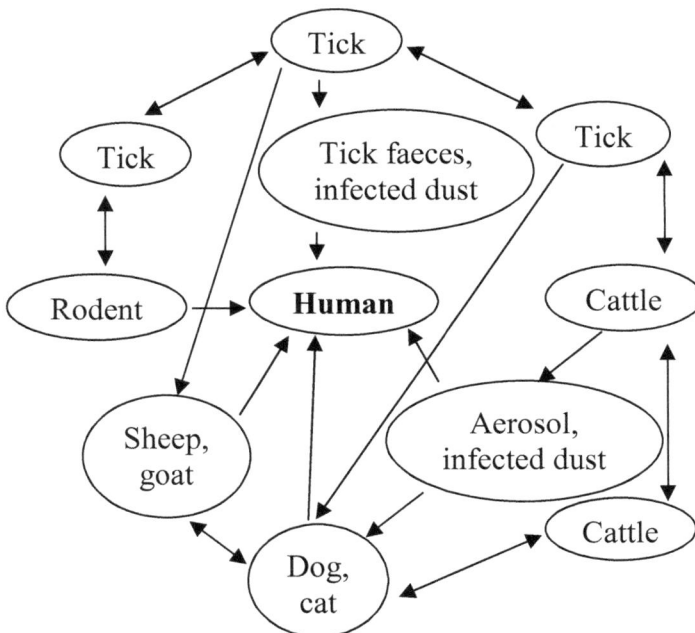

Figure 21.6: Infection Cycle of Q Fever.

3. **Contact:** Humans may be infected from direct or indirect contact with infected animals or their dried excreta. The organism may gain entry into the body though conjunctiva and abraded skin.

4. **Tick bite:** Pigs, dogs, cats and other stray animals can be infected through tick bites.

Disease in Animals

The overt disease is rare in the animals. The disease may cause abortion in some animals like cattle and sheep with subsequent excretion of large number of organisms in the lochia. The recent study also showed that the organism has been associated with metritis, retention of placenta and infertility. The organism propagates in the phagolysosomes, respiratory epithelium and endothelium of blood vessels.

Disease in Humans

The incubation period of disease ranges from 2 to 4 weeks. Infection with C. *burnetii* may lead to inapparent, acute or chronic disease in humans. The acute symptoms usually develop within 2-3 weeks of exposure, although as many as half of humans infected with *C. burnetii* do not show clinical symptoms. The acute form of disease is characterized by high fevers, severe headache, general malaise, myalgia, chills, sweats, non-productive cough, nausea, vomiting, diarrhea, abdominal pain, chest pain *etc.* Although most persons with acute Q fever infection recover, others may experience serious illness with complications that may include pneumonia, granulomatous hepatitis, myocarditis and central nervous system complications. Infected pregnant women may be at risk of abortion or miscarriage. The estimated case fatality rate is less than 2 per cent of hospitalized patients. Treatment with the correct antibiotic may shorten the course of illness for acute Q fever.

Chronic Q fever is a severe disease occurring in less than 5 per cent of acutely infected patients. It may present soon (within 6 weeks) after an acute infection, or may manifest years later. The three groups at highest risk for chronic Q fever are pregnant women, immunosuppressed persons and patients with a pre-existing heart valve defects. Endocarditis is the major form of chronic disease, comprising 60-70 per cent of all reported cases. The estimated case fatality rate in untreated patients with endocarditis is 25-60 per cent. Patients with endocarditis require early diagnosis and long-term antibiotic treatment (at least 18 months) for a successful outcome. Other forms of chronic Q fever include aortic aneurysms and infections of the bone, liver or reproductive organs, such as the testes in males. *C. burnetii* has the ability to persist for long periods of time in the host after infection. Although the majority of people with acute Q fever recover completely. A post-Q fever fatigue syndrome has been reported in 10-25 per cent of some acute patients. This syndrome is characterized by constant or recurring fatigue, night sweats, severe headaches, photophobia, myalgia, arthralgia and difficulty in sleeping.

Diagnosis

1. **Clinical examination:** There are several aspects of Q fever that make it

challenging for healthcare providers to diagnose and treat. The symptoms vary from patient to patient and can be difficult to distinguish from other diseases. However, this can be diagnosed on the basis of history and the triad of high irregular fever, retrobulbar headache and atypical pneumonia without catarrhal symptoms.

2. **Serological examination:** It can be done by using immunofluorescent assay or compliment fixation test. At least fourfold change in the titre between two serum samples in 3 to 6 weeks is suggestive of confirmatory diagnosis.

3. **Cultural examination:** The disease can be diagnosed on the basis of isolation and identification of organisms from the buffy coat of blood or from biopsy material in cell cultures or embryonated eggs. Cultural diagnosis is recommended only in reference laboratories due to high risk of infection with this method.

4. **Molecular technique:** The PCR technique is helpful in the detection of organism from samples.

Note: Differential diagnosis must be made with ornithosis, mycoplasmal and other atypical pneumonia, influenza, salmonellosis, leptospirosis *etc.*

Disease Management in Animals

1. **Restriction on movement:** Restriction on access to barns and laboratories may be helpful in prevention of spread of infection.

2. **Vaccination:** Vaccination with vaccine prepared from inactivated phase 1 *Coxiella burnetii* is effective in the prevention of disease.

3. **Treatment:** The disease can be treated with tetracycline or other antibiotics.

4. **Hygienic measures:** These include proper disposal of placenta, birth products, fetal membranes and aborted fetuses. Use appropriate procedures for bagging, autoclaving, and washing of laboratory clothing are also important in restricting the spread of disease.

5. **Quarantine measure:** It can be adopted for imported animals to prevent the spread of disease.

6. **Miscellaneous measures:** These may include vector control, environmental sanitation and proper disposal of placenta from infected animals for prevention of spread of disease. It is also important to ensure that holding facilities for sheep should be located away from populated areas. Animals should be routinely tested for antibodies to *C. burnetii*.

Disease Management in Humans

1. **Hygienic measures:** Strict hygienic measures must be adopted at the animal farms.

2. **Heat treatment of milk and milk products:** Milk and milk products should be adequately heat treated before consumption in order to prevent the spread of disease.

3. **Health education:** Public should be educated about source of infection and mode of spread of disease. Health education to the persons at highest risk for developing chronic Q fever, especially persons with pre-existing cardiac valvular disease or individuals with vascular grafts is also important.

4. **Vaccination:** Vaccination with vaccine containing inactivated phase 1 C. *burnetii* is useful in high risk of acquiring the infection. A 0.5 ml vaccine is administered subcutaneously in the upper arm. The immunity is lost for at least 5 years.

5. **Treatment:** Treatment is more likely to be effective if started in the first three days of symptoms. Early treatment with doxycycline @ 100 mg, orally, daily for 2-3 weeks is effective. Tetracycline is an alternative drug that can be used @ 500 mg, orally, four times a day. Ofloxacin @ 600 mg or pefloxacin @ 800 mg orally, daily for 2-3 weeks have also been found effective. The chronic Q fever requires treatment for long period. A combination of doxycycline @ 100 mg, orally, twice a day and ofloxacin @ 200 mg, orally, three times a day is recommended for the treatment of chronic Q fever.

Chapter 22

Staphylococcal Infections

Staphylococci cause several infections in animals and humans. Staphylococci cause foodborne disease, skin infections, abscess *etc.*

Etiology

Zoonotic infections in humans are caused by *Staphylococcus aureus* and *Staphylococcus intermedius.*

Epidemiology

Staphylococcal infections are reported worldwide in animals and humans. Staphylococci are found in several animal species. They are commensal organisms on skin and mucous membranes. Butchers, meat handlers and meat industry workers are at higher risk of acquiring infections.

Transmission

Infections are transmitted through contact of skin lesions with tissues, skin and bones of infected animals. Infections are also transmitted by infected animal bites and scratches.

Disease in Animals

Staphylococci are one of the most important causes of mastitis in domestic animals. Staphylococci cause skin infections in cattle, sheep, horse and pig. Staphylococcal dermatitis is observed in sheep, swine, horse and mink. The organism also causes septicemia in pig, pigeon and duck. *Staphylococcus* spp. causes abscess formation in sheep and rabbits.

Disease in Humans

Immunocompetent persons are more susceptible to staphylococcal infections.

In these persons staphylococci penetrate the skin or mucous membranes and cause purulent local infections such as localized impetigo, folliculitis and formation of abscess, furuncles and carbuncles. Generalized pyoderma may develop in some persons. Complications may lead to sinusitis, osteomyelitis and otitis media.

Diagnosis

1. **Cultural examination:** Staphylococci can be isolated from faeces, urine, blood, pus and other exudates using selective media such as mannitol salt agar (MSA), Baird-Parker's egg-yolk tellurite agar and *Staphylococcus aureus* medium No. 110. On mannitol salt ager, *Staphylococcus aureus* forms 1 mm diameter yellow colonies surrounded by the yellow medium due to acid formation. Sodium chloride present in the medium inhibits the growth of most other bacteria but *Staphylococcus aureus* is tolerant of sodium chloride. On Baird-Parker agar, the organism forms 1-2 mm diameter shiny jet-black colonies.

2. **Morphological examination:** On Gram's staining; staphylococci exhibit as Gram-positive cocci with arrangement of grape-like clusters.

3. **Serological examination:** This is useful for diagnosis of systemic infection such as staphylococcal endocarditis. However, serological diagnosis of localized infections caused by staphylococci is not reliable.

4. **Molecular technique:** PCR based DNA fingerprinting is useful for diagnosis of staphylococcal infections. Subspecies typing of staphylococci can be done by molecular method like PFGE (Pulsed-Field Gel Electrophoresis).

Disease Management in Animals

1. **Hygienic measures:** High standards of hygiene should be adopted at animal farms.

2. **Treatment:** Staphylococcal infections can be treated with penicillin.

Disease Management in Humans

1. **Hygienic measures:** Hygienic measures should be adopted while coming in contact with animals.

2. **Protective measures:** Persons at risk should wear the gloves when working with animal tissues.

3. **Treatment:** Staphylococcal infections can be treated with penicillin. However, in recent years the problems of antibiotic resistance in staphylococci have increased. Therefore, antibiotic sensitivity testing is important before its use for treatment.

Chapter 23

Streptococcal Infections

There are several serogroups of *Streptococcus* spp. Some are of zoonotic importance. Serogroup A (*Streptococcus pyogenes*) is a common agent of human infections that can live in a person's nose and throat. Animals are not common reservoirs for this organism. Organism can occasionally be isolated from cattle, monkey and ducks. Serogroup B streptococci (*Streptococcus agalactiae*) do not have too much zoonotic importance because animals and human strains differ in serotype. The organisms of zoonotic importance are serogroup C streptococci (*Streptococcus equi*), serogroup G, L (*Streptococcus dysgalactiae*) and serogroup R streptococci (*Streptococcus suis*).

Streptococcus equi Infections

In humans, the infections with *Streptococcus equi* contracted from animals, usually occur in sporadic form.

Etiology

Streptococcus equi subsp. *zooepidemicus* (formerly *Streptococcus zooepidemicus*) more frequently causes zoonotic disease than *Streptococcus equi* subsp. *equi*.

Epidemiology

Streptococcus equi subsp. *equi* is found in the nasopharynx, upper respiratory tract, on the tonsils, mucous membranes of healthy horses and cattle. The disease is enzootic in animals. Animal keepers, animal dealers, farmers and persons in frequent contact with horses and other domestic animals are at risk of acquiring infections with this organism.

Transmission

Transmission occurs mainly through direct contact with animals that excrete large number of pathogens. Contact with nasal secretions, bite or by consumption of raw infected/contaminated milk and milk products.

Disease in Animals

Streptococcus equi causes mastitis, arthritis, endometritis, purulent wound infections and septicemia in the animals.

Disease in Humans

The disease is characterized by wound infection (impetigo), upper respiratory tract infection, pneumonia, pleuritis, endocarditis, meningitis, arthritis, lymphadenopathy and septicemia. There may be complication such as glomerulonephritis in the later stage of disease.

Diagnosis

1. **Cultural examination:** Organism can be isolated on conventional media enriched with blood. On blood agar, streptococci of Lancefield's group C and G produce wide zone of ß-hemolytic colonies.

2. **Biochemical examination:** Organism can be identified by the characteristic of fermentation of carbohydrates like sorbitol and trehalose.

3. **Serological examination:** Organism can be identified by serogrouping (organism belongs to the Lancefield serogroup C).

4. **Molecular technique:** DNA finger printing is used for epidemiological purpose.

Disease Management in Animals

1. **Quarantine measures:** Quarantine measures are important in prevention of infection in equines. Stop all movement of horses on and off the affected premises immediately and until further notice.

2. **Hygienic measures:** In animals the disease can be managed by adopting strict hygienic measures in the animal houses. Following removal of organic material from stables, all surfaces should be thoroughly soaked in an appropriate liquid disinfectant or steam treated and allowed to dry. Manure and waste feed from infected animals should be composted in an isolated location. Pastures used to hold infected animals should be rested for four weeks. Care should be taken to disinfect water troughs at least once daily during an outbreak. Horse vans should be hosed clean and disinfected after each use.

3. **Therapeutic measures:** The disease can be treated by using antibiotic like penicillin.

Disease Management in Humans

1. Avoid contact with infected animals.
2. Avoid consumption of raw milk.

3. Follow good hygiene practices.
4. Treatment with antibiotic like penicillin G (@ 2-5 million IU, i.v., four times a day).

Streptococcus suis Infections

Streptococcus suis infections involve meningitis other neural problems.

Etiology

Streptococcus suis has several serovars. Serovar 2 is the most important agent of zoonoses.

Epidemiology

The disease has been reported from UK, US and New Zealand. The main reservoirs of *Streptococcus suis* are pigs. The disease is most commonly found in piglets particularly in overcrowding conditions. The organism may persist for long time in clinically healthy animals particularly in the tonsils.

Transmission

The disease is transmitted by direct contact with infected pigs, pork and via conjunctiva. Wound infection may occur through contaminated knife. Infection can also be transmitted by consumption of raw or improperly cooked pork.

Disease in Humans

The incubation period of disease is very short that is few hours to 2 days. The disease is characterized by meningitis and involvement of eighth nerve which leads to loss of hearing and balance. Deafness may occur in the later stage of disease. *Streptococcus suis* serovar 2 does not cause meningitis. This serovar has been incriminated as a cause of nystagmus, uveitis, endophthalmitis, endocarditis, myocarditis, arthritis and pneumonia.

Diagnosis

1. **Clinical examination:** A history of contact with pigs or pork and symptoms associated with meningitis and involvement of eighth nerve is highly suggestive of infection.
2. **Cultural examination:** Confirmatory diagnosis can be made by isolation of the organism from blood, spinal fluid and joint fluid on sheep blood agar. The organism shows α-hemolytic colonies on blood agar medium.
3. **Biochemical examination:** The organism can be identified by the characteristic of fermentation of carbohydrate like trehalose, esculin hydrolysis and positive Voges-Proskauer test.
4. **Serological examination:** The organism can be identified by serogrouping (organism belongs to the Lancefield serogroup R, S or T).

Disease Management in Humans

1. **Personal protection:** Persons in contact with infected pigs or pork should wear the protective clothing and gloves.

2. **Management of wound:** Wound should be disinfected and prophylactically treated with penicillin.

3. **Treatment:** Penicillin is the drug of choice for treatment of this infection. Penicillin can be given @ 2-5 million IU, i.v., four times a day. Alternative drugs such as ampicillin or chloramphenicol in combination with aminoglycosides can also be used.

Chapter 24

Tuberculosis

Synonyms

Pearl's disease, Pott's disease, Scrofula, Great white plague, Great white scourge

Tuberculosis (TB) is a chronic contagious disease of animals and humans. The disease is characterized by development of tubercle followed by caseation and calcification. Tuberculosis most commonly affects the lungs.

Etiology

Tuberculosis is caused by many species of *Mycobacterium*. The organism is non-spore forming and non motile. Bovine tuberculosis is caused by *Mycobacterium bovis*; however, cattle are also infected from human type (*Mycobacterium tuberculosis*). Human tuberculosis is mainly caused by *M. tuberculosis* and rarely by *M. bovis*, *M. africanum*, *M. microti* and *M. canettii*. These pathogens constitute the *Mycobacterium tuberculosis* complex. The principal agent of zoonotic tuberculosis is *Mycobacterium bovis*.

Epidemiology

Tuberculosis occurs worldwide. The disease is highly prevalent in tropical and subtropical countries. It is an important occupational zoonotic disease. Farmers, animal handlers, abattoir and laboratory workers are at higher risks of infection. In India, the disease is endemic and has been reported from all the states. The disease has been eradicated from Canada, Denmark, Finland, Germany, Great Britain, Switzerland, USA and Russia by application of tuberculin test followed by removal of positive reactors. In human beings the disease is more prevalent in males than in females. Human beings do not have inherited immunity against tuberculosis. Tuberculosis is an opportunistic infection that most frequently kills the HIV infected persons. The people infected with both TB and HIV is 25-30 times more likely to develop tuberculosis than people infected only with tuberculosis. Reoccurrence of tuberculosis is more in HIV infected persons.

Tuberculosis is a top infectious disease killer worldwide. In 2014, 9.6 million people fell ill with TB and 1.5 million died from this disease. Over 95 per cent of TB deaths occur in low- and middle-income countries, and it is among the top 5 causes of death for women aged 15 to 44. In 2014, an estimated 1 million children became ill with TB and 140 000 children died of TB. TB is a leading killer of HIV-positive people. In 2015, 1 in 3 HIV deaths was due to TB.

Globally in 2014, an estimated 480 000 people developed multidrug-resistant TB (MDR-TB). TB incidence has fallen by an average of 1.5 per cent per year since 2000 and is now 18 per cent lower than the level of 2000. The TB death rate dropped 47 per cent between 1990 and 2015. An estimated 43 million lives were saved through TB diagnosis and treatment between 2000 and 2014.

Transmission

1. **Inhalation:** Tuberculosis is mainly transmitted through inhalation of contaminated air or infected aerosols.

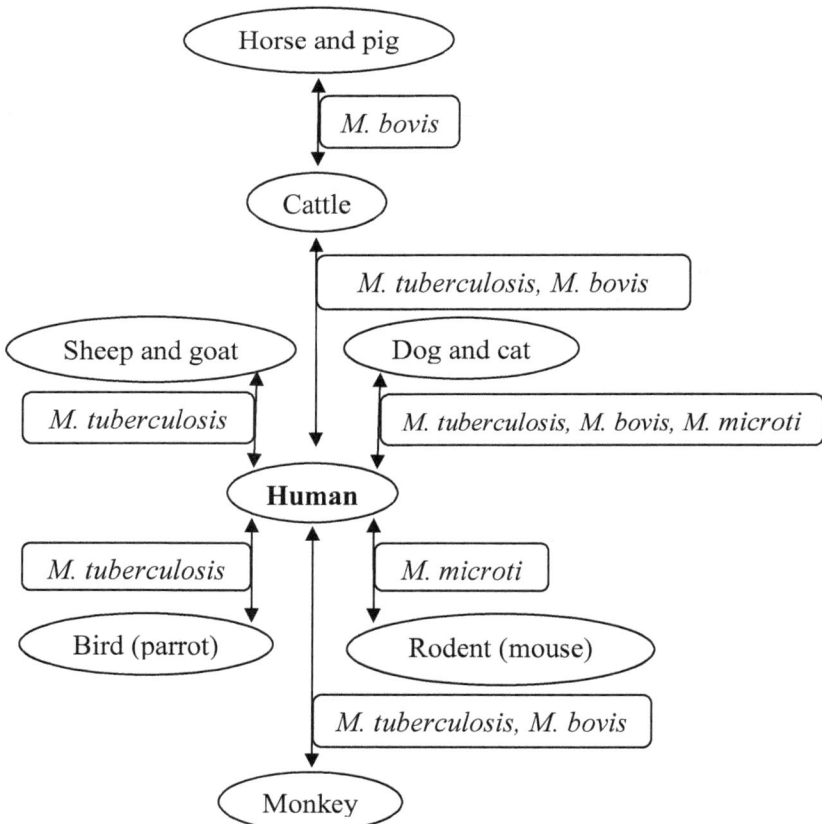

Figure 24.1: Infection Cycle of *M. tuberculosis* Complex.

2. **Ingestion:** It is also important route of transmission. Animals get the infection through feed, water and milk contaminated with sputum, faeces, urine, vaginal discharges, semen *etc.* Calf may get the infection through infected milk or through the umbilical vein. Carnivores are oftenly infected from meat and milk from infected animals.

Disease in Animals

In cattle the respiratory system is most commonly affected. It is characterized by dyspnea, increased respiration rate and hacking cough. The involvement of digestive system is characterized by obstinate diarrhea, chronic bloat and abdominal pain. The involvement of uterus is characterized by purulent and caseous discharge tinged with blood. The udder infection is characterized by formation of small nodules in the mammary tissues and enlargement of supramammary lymph nodes. Milliary TB is characterized by multiple lesions of varying size may involve pleura and peritoneum.

Disease in Humans

The disease primarily affects the lungs and cause pulmonary TB. It can also affect intestine, meninges, bones and joints, skin and other tissues of the body. The disease, particularly pulmonary form is characterized by continuous fever, chest pain, hemoptysis and persistent cough.

Diagnosis in Animals

Disease diagnosis can be made on the basis of clinical signs, clinical pathology (examination of sputum and other discharges), guinea pig inoculation, radiological examination and tuberculin test.

Diagnosis in Humans

1. **Tuberculin test:** As the diagnostic value of tuberculin test is invalidated, this test has little value as a case finding tool.

2. **Radiological examination:** The WHO expert committee on tuberculosis concluded that indiscriminate use of this technique as a case finding tool should be abandoned. This method has high cost, high proportion of erroneous interpretation of X-ray films, lack of definitiveness that is the mere presence of X-ray shadows is not indicative of a case unless the presence of tubercle bacilli are demonstrated.

3. **Cultural examination:** International Union against tuberculosis (IUT) medium is a standard medium for the isolation of mycobacteria. This is a modification of the original Lowenstein-Jensen glycerol egg medium. Colonies of *M. tuberculosis* are of an off-white (buff) colour and dry breadcrumb-like appearance.

4. **Sputum examination:** Sputum examination by direct microscopy is now considered the method of choice. The reliability, cheapness and ease of direct microscopic examination have made this method number one case finding method all over the world.

Disease Management in Animals

1. **Hygienic measure:** Premises inhabited by sick animals should be properly disinfected. Feeding trough, watering, utensils and milking pans should be thoroughly scrubbed and disinfected.

2. **Test and slaughter:** Eradication of disease in animals has been achieved in developed countries by adopting test and slaughter policy.

3. **Vaccination:** BCG vaccine can be used in calves but may create problems in the vaccinated herds.

4. **Treatment:** Isoniazid or streptomycin can be used in the treatment of tuberculosis. Isoniazid (@ 5 mg per kg body weight for 45 days) can be used. Streptomycin can be used as an alternative drug (@ 11 mg per kg body weight, i.m. for 45 days).

Disease Management in Humans

1. **Vaccination:** BCG (Bacille Calmette-Guerin) vaccine is a living bovine strain of tubercle bacillus which has been rendered avirulent. The WHO has recommended the "Danish 1331" strain for the production of BCG vaccine. BCG vaccine containing 0.1 mg in 0.1 ml is injected intradermally with the help of tuberculin syringe just above the insertion of the deltoid muscles. Booster doses of BCG are not recommended by the WHO. BCG vaccine is contraindicated for individuals with severely impaired immunity and individuals with HIV infection.

2. **Treatment:** Chemotherapy is indicated in every case of active tuberculosis. The object of treatment is to cure that is the elimination of both the fast and slowly multiplying bacilli including the persisters from the patient's body. There are several antituberculosis drugs for the treatment with different strategies.

    i. **Long course regimen:** The WHO expert committee on "tuberculosis control" affirmed that the long course regimen still remain basic requirements for use in many developing countries. This regimen may be of two types-

    Daily regimen: The most commonly used combination in India is hydrazide of isonicotinic acid (INH) @ 300 mg for adult in a combination with thioacetazone @ 150 mg given together in a single daily dose for 18 months. This method is suitable for rural areas where medical facilities are not readily available.

    Bi-weekly regimen: Streptomycin @ 1 g in a combination with INH @ 600 mg and pyridoxine (to prevent polyneuritis) @ 10 mg is given in standard bi-weekly regimen.

    ii. **Short course regimen:** The advantage of this course is lower failure rates and reduction in frequency of emergence of drug resistant bacilli. There are number of short course regimens of 6 months. Short course regimens include initial intensive phase with four drugs

(INH, rafampicin, pyrazinamide, supplemented by streptomycin or ethambutol) for 2 months followed by continuation phase with two drugs (INH and rafampicin or thioacetazone) for 4 months daily or intermittently.

Directly observed treatment short course (DOTS) chemotherapy: This short course regimen has been effective and being used worldwide. In initial intensive phase of DOTS chemotherapy, the drug is taken/ swallowed by patient in the presence of medical staff. In continuation phase, the drug is issued to patient for one week in a multibilister combipack and the first dose is swallowed by patient in the presence of medical staff. The drugs are further issued to the patients for one week after returning of used multibilister combipack. DOTS chemotherapy is also effective in HIV cases of tuberculosis if timely diagnosed and properly treated.

3. **Miscellaneous measure:** Strict hygienic conditions must be maintained particularly by infected persons. Sputum/mucous of patient must be disinfected and disposed of properly. Infected persons should not be allowed for milking of animals as well as handling of common foods. Milk must be adequately heat treated prior to its consumption.

Chapter 25

Tularemia

Synonyms

Rabbit fever, Deer fly fever, Ohara's disease

Tularemia is a bacterial zoonotic disease of the Northern hemisphere. The bacterium (*Francisella tularensis*) is highly virulent for humans and a range of animals such as rodents, hares and rabbits. It may cause epidemics and epizootics.

Etiology

F. tularensis subspecies *tularensis* (type A) is one of the most infectious pathogens known in humans. It is an intracellular organism. The highly virulent *F. tularensis tularensis* (type A) is found in the USA, whereas the less virulent *F. tularensis holarctica* (type B) is found throughout the Northern hemisphere.

Epidemiology

Tularemia is reported from most countries in the Northern hemisphere, although its occurrence varies widely from one region to another. The disease has been reported from Australia, Eastern and Western Europe, Portugal, Spain, Sweden and UN Administered Province of Kosovo (Serbia), USA, Canada, Mexico, Japan and China. In an endemic area, tularemia may occur annually within a 5-year period, but may also be absent for more than a decade. The reasons for this temporal variation in the occurrence of outbreaks are not well understood. *F. tularensis* is a potential bioterrorism agent.

Type B tularemia occurs in Eurasia and in North America. In contrast to type A, type B is mainly associated with streams, ponds, lakes, rivers, and semi-aquatic animals such as muskrats and beavers. Tularemia outbreaks have been reported in hares, prairie dogs, mink, ground squirrels, rabbits, hares, beavers, muskrats and in particular, rodents such as meadow voles and water voles. The disease occurs in all age groups. The incidence of the disease is higher in males. The occurrence of the disease is seasonal in countries where the disease is endemic. Higher incidence of the disease occurs during late spring, early autumn and the summer months. Outbreaks of the disease in humans often follow outbreaks of tularemia in rodents.

Transmission

1. **Contact:** Humans are infected most frequently through direct contact with excreta, blood, tissue or body fluid of infected animals.
2. **Ingestion:** Transmission of infection is possible through consumption of contaminated food or water.
3. **Inhalation:** Transmission of infection also occurs through inhalation of dusts contaminated with excreta of infected rodents. Contaminated dusts may be generated particularly during harvesting of grains.
4. **Bites:** Tularemia is transmitted to humans by arthropod (ticks, flies, lice and fleas) bites. It can also be transmitted following bites of cats and squirrels.

Note: There is no human-to-human transmission. Humans are infected by animals only.

Disease in Animals

The disease may occur in acute or chronic form. Acute form is characterized by fever and lethal hemorrhagic septicemia. The chronic form of disease is characterized by formation of abscess in the liver and spleen. Signs and symptoms of tularemia in wild animals are not well documented and are mostly based on post-mortem examinations. The most common finding upon necropsy is an enlarged spleen and pinpoints white necrotic lesions in the spleen and liver. The best-documented clinical cases are in domestic cats and dogs, captive monkeys, prairie dogs and laboratory animals.

Disease in Humans

The incubation period ranges from 1 to 14 days. An acute form of disease is characterized by fever, chills, headache and pronounced lassitude. Depending upon the portal entry, the disease may occur in the following forms:

1. **Ulcero-glandular tularemia:** This form is characterized by development of red papule which enlarges, becomes purulent and ulcerates. Later on the regional lymph nodes become enlarged, suppurate and ulcerate.
2. **Oculoglandular tularemia (Parinaud's conjunctivitis):** If the organism enters through conjunctiva, the syndrome is called Parinaud's conjunctivitis or oculoglandular tularemia. This may be acquired by touching the eye with a contaminated fingers or dust. This form comprises less than 1 per cent of all human cases of tularemia. This form is characterized by fever, unilateral conjunctivitis with granulomatous lesions on the palpebral conjunctiva, swelling of the eyelids, excessive lacrimation, mucopurulent discharge and photophobia.

3. **Oropharyngeal tularemia:** Oropharyngeal tularemia may occur due to ingestion of contaminated water or food. This form is characterized by ulcerative-exudative stomatitis and pharyngitis, with or without tonsillar involvement. Other symptoms of the disease include redness and pustular changes in the mouth and pharyngeal mucous membranes and excessive regional neck lymphadenitis.

4. **Respiratory tularaemia:** This form of tularaemia is contracted by inhalation of contaminated aerosols. The disease occurs most frequently during farming activities. Respiratory tularaemia is characterized by high fever, pneumonia, cough, chest pain, an increased respiratory rate, nausea and vomiting.

5. **Typhoidal tularaemia:** Historically, the typhoidal form was defined as tularaemia devoid of skin or mucous membrane lesion and/or a remarkable lymph node enlargement. A continued use of the term "typhoidal" seems unjustified and confusing. Only when no route of infection can be established may the term still be acceptable.

6. **Childhood tularaemia:** The disease is characterized by lymphadenopathy which occurs more frequently in children than in adults, with a more frequent cervical localization and a tendency for delayed suppuration. Severe pulmonary symptoms also occur more frequently in children than adult.

Diagnosis

1. **Clinical examination:** The clinical diagnosis can be made from history of contact with animals and clinical symptoms.

2. **Cultural examination:** The disease can be diagnosed by isolation of organism from various lesions, pus, lymph node biopsy, specimens, conjunctival secretions, sputum, gastric aspirate or blood using special medium like blood-glucose-cystine agar. Minute droplet-like colonies develop in 72 hours.

3. **Serological examination:** The disease can be diagnosed by tube agglutination test. ELISA is useful for detection of IgM, IgA and IgG antibodies.

4. **Molecular examination:** PCR is a reliable test for diagnosis of tularemia.

Disease Management in Animals

1. **Protective measures:** Avoid feeding of raw rabbit meat especially to felines.

2. **Control of vectors:** It is important to control the vectors like ticks, lice *etc.* in order to prevent the transmission of the disease.

Disease Management in Humans

1. **Vaccination:** Vaccination may be carried out with an attenuated strain

of *F. tularensis*. Vaccination of the human population against tularaemia in endemic regions may be useful; however, currently the vaccine is not available.

2. **Treatment:** The drug of choice is penicillin (@ 10 mg/kg body weight, twice a day i.m.) or gentamicin (@ 1.7 mg/kg body weight, thrice a day) for 1-2 weeks. Ciprofloxacin (@ 750 mg twice a day, orally for 2 weeks) can be used as an alternative drug.

3. **Protective measures:** Protect the water sources from contamination through contact with animals such as rats, mice *etc*. Avoid the hunting of hares and rabbits and consumption of their meat. Avoid exposure to blood-sucking arthropods by wearing long-sleeved clothing, and using mosquito nets or insect repellents.

4. **Hygienic measures:** Avoid consumption of contaminated foods and drinking water. Protect food stores from contact with animals using traps for mice, rats or other rodents. Wash hands after contact with wild and domestic animals.

5. **Other measures:** Regularly inspect domestic animals for signs of disease. In outbreak situations, avoid close contact with domestic animals such as dogs and cats. Avoid dust and aerosols (especially relevant for farmers and landscapers) by closing doors and avoid rooms where aerosols or dust are generated. Avoid exposure to contaminated dusts or aerosols using respiratory masks (protection class FFP3).

Chapter 26

Foodborne Bacterial Diseases

Foodborne diseases encompass a wide spectrum of illnesses and are a growing public health problem worldwide. They are the result of ingestion of foodstuffs contaminated with microorganisms or chemicals. The contamination of food may occur at any stage in the process from food production to consumption ("farm to fork") and can result from environmental contamination, including pollution of water, soil or air. The most common clinical presentation of foodborne disease takes the form of gastrointestinal symptoms; however, such diseases can also have neurological, gynecological, immunological and other symptoms. Multi-organ failure and even cancer may result from ingestion of contaminated foodstuffs. Therefore, foodborne diseases represent a considerable burden of disability as well as mortality.

Foodborne diseases cause diarrheal illness worldwide which results in morbidity and mortality. There is a strong need to strengthen surveillance systems for foodborne diseases. WHO has developed a comprehensive strategy on strengthening foodborne disease surveillance. One of the initiatives to strengthen surveillance systems for foodborne disease was the establishment, in January 2000, of the Global Foodborne Infections Network (GFN), consisting of institutions and individuals working in human health as well as veterinary and food-related disciplines. WHO assists Member States in building capacity to prevent, detect and manage foodborne risks. WHO's activities include generating baseline and trend data on foodborne diseases and supporting implementation of adequate infrastructures like laboratories. WHO is also involved in outbreak investigation and response, pre-harvest control strategies, burden of foodborne disease illness and antimicrobial resistance due to non-human antimicrobial usage. Moreover, WHO has formulated 10 "Golden Rules" for prevention of food poisoning as follows:

1. Food processed for safety.
2. Thoroughly cooked.

3. Eat immediately.
4. Store carefully.
5. Reheat thoroughly.
6. No contact between raw and cooked food.
7. Wash hands.
8. Keep food preparation surface clean.
9. Protect from pests.
10. Use potable water.

Some important foodborne bacterial diseases are discussed as follows:

Aeromonas hydrophila Infection

Etiology

Aeromonas hydrophila is a Gram-negative, rod-shaped and motile organism. It belongs to the family Aeromonadaceae.

Epidemiology

Aeromonas is ubiquitous in environment. It is present in all type of water worldwide as well as food and soil. It is highly prevalent in fish, shellfish, reptiles and amphibians. *Aeromonas* infections have been reported in cattle, buffalo, poultry and humans.

Transmission

The disease is transmitted due to consumption of contaminated seafood, chicken, red meat and water.

Disease in Humans

Aeromonas hydrophila causes gastroenteritis which resembles dysentery-like disease as observed in case of shigellosis. It causes watery diarrhea due to its cytotonic enterotoxin.

Diagnosis

Aeromonas is similar to bacteria in the coliform group and can be isolated from similar environment. Ampicillin dextrin agar (ADA) is used for isolation of *Aeromonas* spp. Colonies spread with Nadi reagent (1 per cent solution of N, N, N, N'- tetra methyl p- phenylene- diammonium chloride). A positive Nadi reaction (dextrin degradation) is indicated by a purple colour at the periphery of the colony. Dextrin fermentation is also indicated by yellow colonies. *Aeromonas* spp. appears as large convex yellow colonies with purple periphery. On blood agar most strains of *Aeromonas* give wide zone of haemolysis, but strains of *Aeromonas caviae* are usually non-haemolytic.

Disease Management in Humans

Avoid consumption of contaminated foods. Raw seafood, meat and milk should be adequately heat treated in order to prevent the spread of infection.

Bacillus cereus Food Poisoning

Bacillus cereus has been recognized as a cause of food poisoning with increasing frequency in recent years.

Etiology

The disease is caused by *Bacillus cereus* which is a Gram-positive, aerobic, spore forming and motile rod. The organism is ubiquitous in soil and in raw dried and processed foods. The spores can survive cooking temperature and germinate and multiply rapidly when the food is held at favourable temperature.

Epidemiology

The disease occurs worldwide and constitutes about 0.8-22.4 per cent of all the foodborne diseases. The organism has been isolated from a variety of foods of agricultural and animal origin.

Transmission

The disease is transmitted by ingestion of contaminated boiled rice (dried and stored), meat and vegetables.

Disease in Humans

The disease is found in two distinct forms on the basis of toxin production by the organism. The toxins are preformed and stable.

1. **Emetic form:** The incubation period is about 1-6 hrs and characterized by predominantly upper gastrointestinal tract symptoms rather like staphylococcal food poisoning.

2. **Diarrheal form:** The incubation period is comparatively longer (12-24 hrs). This form is characterized by predominantly lower intestinal tract symptoms like *Cl. perfringens* food poisoning (diarrhea, abdominal pain, nausea with little or no vomiting and no fever).

Diagnosis

1. **Cultural examination:** Confirmatory diagnosis can be made by isolation of *B. cereus* from epidemiologically incriminated food (1 lac or more organisms per g). Mannitol egg-yolk polymyxin (MEYP) agar, polymyxin pyruvate egg-yolk mannitol bromothymol blue agar (PEMBA) and phenol red egg-yolk polymyxin agar are selective media for isolation of the organism. Polymyxin present in the medium inhibits Gram-negative bacteria. The diagnostic features of the medium rely upon the failure of *B. cereus* to utilize mannitol and the ability of most strains to produce phospholipase C. On MEYP and PEMBA, *B. cereus* forms gray-white flat

or slightly raised colonies with an opaque zone of turbidity. On phenol red egg-yolk polymyxin agar, it forms opaque and pink colour colonies.

2. **Animal inoculation test:** *B. cereus* causes accumulation of fluid in the rabbit ligated gut segment, induce a vascular permeability reaction on the rabbit skin and is lethal to mouse.

Disease Management in Humans

1. **Treatment:** The treatment is symptomatic.

2. **Proper storage of food:** The disease may be easily prevented by cooling and storage of food. Cooked rice should not be stored for long periods above 10°C.

3. **Hygienic measures:** Strict hygienic measures during preparation, handling and storage of food are helpful in prevention of disease.

Botulism

Botulism is a toxemic, paralytic disease of animals, birds and humans, caused by ingestion of toxins of *Clostridium botulinum*, which is a soil saprophyte and common inhabitant of digestive tract of animals.

Etiology

Human botulism is a serious but relatively rare paralytic disease, caused by one of the most potent toxins that exist. The disease is caused by *Cl. Botulinum*, a Gram-positive rod-shaped, anaerobic and endospore-forming organism. The organism proliferates in decaying organic matter. The spores of the organism are highly resistant and can survive for a long period in decaying carcases. On the basis of neurotoxin production by different strains, the organism has been classified into seven antigenic types *viz.*, A, B, C, D, E, F and G. The exotoxins of *Cl. botulinum* type A, B, E and rarely F are responsible for food poisoning in the humans.

Epidemiology

The disease occurs worldwide. The disease has been reported from most of the tropical countries. The organism is widely distributed in the soil, dust and intestinal tract of animals. The spores are also present in the faeces of mammals and birds and on animal carcases. The most common foods incriminated in botulism are home-canned vegetables, home-made cheese and smoked or pickled fish. An outbreak of botulism occurred in 2006 in northern Thailand, which affected 152 people without causing any deaths, was caused due to consumption of home-preserved bamboo shoots at a festive gathering.

Transmission

The disease is transmitted by ingestion of contaminated food and water. The organism is saprophytic and produces toxin in the decaying organic matter or soil. The organism does not produce toxin within the animal's body but preformed toxins are the main source of disease transmission. Human-to-human transmission of botulism does not occur.

Disease in Animals

The disease may occur in per-acute, acute, sub-acute or chronic form. The disease is characterized by profuse salivation, partial or complete paralysis of muscles of locomotion, inability to chew and swallow and protrusion of tongue from mouth. There is no rise of body temperature.

Disease in Humans

The incubation period of foodborne botulism in human being is 12-36 hours. This is variable in animals and may be up to 2-6 days depending upon the amount of the toxin ingested. The toxin is preformed in the food (intra-dietetic) under suitable anaerobic conditions. It acts upon the parasympathetic nervous system. Botulism differs from other food poisonings in that the gastrointestinal systems are very slight. The disease is characterized by dysphagia, diplopia, ptosis, dysarthria, muscle weakness, quadriplegia and blurring of vision. The condition is frequently fatal, death occurs 4-8 days later due to respiratory or cardiac failure.

Other kinds of botulism may occur in humans. Wound botulism occurs when wounds are infected with *C. botulinum* that secretes the toxin. Infant botulism occurs typically in children less than 1 year old. This occurs after eating bacterial spores from food that develop into toxins in the intestines.

Diagnosis

1. **Clinical examination:** The disease can be diagnosed on the basis of clinical symptoms.

2. **Cultural examination:** *Cl. botulinum* isolation (CBI) agar, Robertson cooked meat medium (RCM) and Tryptone Peptone Glucose Yeast (TPGY) medium are recommended for isolation of *Cl. botulinum*. Typical colonies are raised, or flat, smooth or rough and commonly show some spreading and have irregular edge.

3. **Molecular examination:** Rapid detection of organism in the food can be performed using PCR technique.

4. **Animal inoculation test:** Inoculation of infected material (food) in the mouse or guinea pig.

5. **Demonstration of toxins:** Demonstration of toxins in food/feed, stomach and intestinal contents, serum, faeces *etc.* may be useful for diagnosis of the disease.

Disease Management in Animals

Cases of botulism can be treated by using antitoxins. Botulism can be best prevented by protecting the foods from its contamination.

Disease Management in Humans

1. Active immunization can be done by using botulinum toxoid to prevent botulism.

2. If a case of botulism is found the antitoxin (@ 50,000 to 1, 00,000 units i.v.) should be given to all individuals partaking of food. Antitoxin is ineffective if the toxin is already fixed to the nervous tissues.

3. To reverse the neuromuscular block, the guanidine hydrochloride can be given (@ 15-40 mg/kg body weight).

4. The food can be rendered safe by heat treatment at 100°C for few minutes.

Campylobacteriosis

Campylobacteriosis is one of the most common and widely distributed foodborne diseases.

Etiology

Campylobacteriosis is caused by *Campylobacter* spp., which is a Gram-negative, spirally curved rod. *Campylobacter* was formerly known as *Vibrio*. *Campylobacter* enteric infection especially in humans is caused by *Campylobacter jejuni* and *Campylobacter coli*. However, *Campylobacter lari* has occasionally been found to be involved. *Campylobacter jejuni* and *Campylobacter lari* are particularly adapted to birds, while *Campylobacter coli* to pigs.

Epidemiology

Campylobacteriosis occurs worldwide in humans and animals. In developed countries, the infection rates have surpassed than that of entero-pathogens like *Salmonella* spp. and *Shigella* spp. In general the prevalence of disease is higher in infants than in the adults.

Transmission

Ingestion of contaminated food products, water and raw milk is the principal mode of transmission in humans. Campylobacteriosis also occurs as an occupational disease involving the workers engaged in processing of food and handling of food animals and their produce.

Disease in Animals

The disease is caused by *Campylobacter jejuni* in animals, manifests mainly the intestinal symptoms such as enteritis, winter dysentery and diarrhea.

Disease in Humans

The disease is characterized by moderate to severe diarrhea. Following an average incubation period of 3-5 days, there is fever accompanied by malaise, headache, dizziness, myalgia and abdominal pain. The stool may be foul-smelling and contain blood and mucous. No clinical disease in humans has been attributed to *Campylobacter fetus* subspecies *venerialis*.

Diagnosis

1. **Cultural examination:** Cultural examination of faeces from the diarrheic cases and asymptomatic carrier and blood from febrile cases of acute

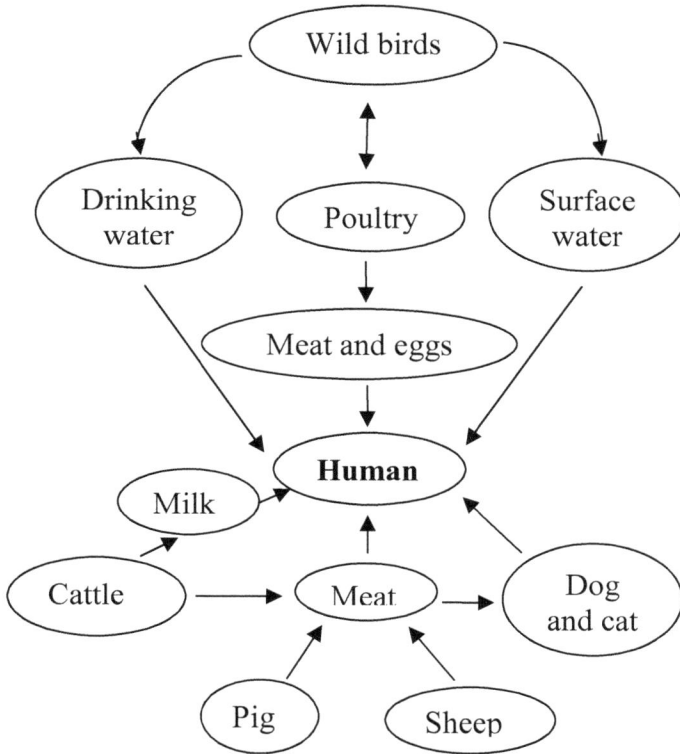

Figure 26.1: Infection Cycle of Campylobacteriosis.

enteritis is important. Commonly used selective media for isolation of *Campylobacter* are Skirrow's selective medium, Butzler's selective medium, Blasser-Wang selective medium and *Campylobacter* blood-free medium (Charcoal cephoperazone deoxycholate agar). *Campylobacter* forms small, flat translucent and effuse colonies on these media. On blood agar medium, it forms non-hemolytic, flat or slightly raised colonies with irregular edges.

2. **Serological tests:** Tests such as agglutination test, compliment fixation test, RIA and ELISA can be used for detection of antibodies in humans.

3. **Molecular techniques:** Molecular techniques such as PCR&DNA probes are helpful in the detection of *Campylobacter* spp.

Disease Management in Animals

1. **Hygienic measures:** Food contamination should be prevented. High standards of hygiene should be adopted on the farms.

2. **Treatment:** Treatment of infection can be done using antibiotics like erythromycin.

Disease Management in Humans

1. **Treatment:** Replacement of fluid and electrolytes is of prime importance. Antibiotic treatment is carried out only when diarrhea is long lasting or recurs. Infection can be treated using antibiotics like erythromycin @ 500 mg twice a day or doxycycline @ 100 mg twice a day, orally for one week.

2. **Hygienic measures:** Enforcement of strict hygiene during slaughter of animals, post-slaughter processing of meat and irradiation of packaged foods are important in preventing the disease.

3. **Miscellaneous measures:** These may include the pasteurization of milk and proper cooking or frying of pork and poultry and adequate chlorination of water.

Clostridium perfringens Gastroenteritis

The organism causes many diseases in animals and foodborne illness in human beings.

Etiology

Clostridium perfringens (formerly called *Clostridium welchi*) is the etiological agent for many diseases in man and animals. It is a Gram-positive, rod-shaped, non motile and endospore forming organism. All warm blooded animals carry *Cl. perfringens* in their intestine. The organism is found in faeces of animals and humans, soil, water and air. The disease outbreaks usually occur due to ingestion of contaminated meat and meat products and poultry. There are many strains of *Cl. perfringens* such as type A, B, C, D and E depending on toxin production. They produce alpha, beta, epsilon and iota toxins. All strains produce alpha toxin which has hemolytic and necrotizing activity.

Epidemiology

The disease occurs worldwide. Animal infection usually occurs due to grazing of pasture contaminated with faeces. Organism has been recovered from specimens of beef, pork *etc.*

Transmission

Human infection occurs due to ingestion of contaminated meat and meat products. Foods are usually contaminated due to unhygienic handling.

Disease in Animals

Cl. perfringens strains cause several diseases. *Cl. perfringens* type A causes gas gangrene in animals. It is characterized by rapid necrosis of muscles and subcutaneous tissues and development of hemoglobin-stained edema containing gas bubbles, local greenish coloration or blackening tissues and profound toxemia resulting in death. *Cl. perfringens* type B causes lamb dysentery. *Cl. perfringens* type C causes struck in sheep (hemorrhagic enterotoxaemia and necrotic and ulcerative changes in the small intestine). *Cl. perfringens* type D causes enterotoxaemia in calves and sheep and pulpy kidney disease (softening of kidneys) in sheep. *Cl. perfringens* type E causes enterotoxaemia in calves.

Disease in Humans

The incubation period ranges from 6-24 hours. The disease is characterized by diarrhea, abdominal cramps and mild or no fever. The duration of illness is short (usually one day). Recovery is rapid. No deaths have been reported.

Diagnosis

1. **Cultural examination:** Cultural examination of incriminated food, faeces and environmental specimens can be made for isolation and identification of organism. Recommended plating media for selective isolation of *Cl. perfringens* are tryptose sulphite cycloserine (TSC) agar, sulphite polymyxin sulphadiazine (SPS) agar and Shahid-Ferguson perfringens (SFP) agar. On (SPS) agar, *Cl. perfringens* reduces the sulphite to sulphide which reacts with iron and forms black iron sulphide precipitate, seen as black colonies. On SFP agar, it produces black colonies surrounded by opaque zone.
2. **Serological test:** ELISA is useful for diagnosis of the disease.
3. **Enterotoxin assay:** It is also helpful in the diagnosis of this disease.
4. **Molecular diagnosis:** Molecular detection of organism can be made by using PCR, DNA probes *etc.*

Disease Management in Animals

1. **Treatment:** Penicillin is effective against *Cl. perfringens* organism.
2. **Hygienic measures:** Environmental sanitation is important for prevention and control of *Cl. perfringens* infections. Pastures should be protected from contamination in order to prevent the spread of disease in the animals.

Disease Management in Humans

1. **Hygienic measures:** High standards of personal hygiene should be adopted during preparation and storage of beef and poultry.
2. **Miscellaneous measures:** Food should be thoroughly cooked. The food must always be cooled immediately after cooking for storage purpose.

Escherichia coli Infection

Synonyms

Colibacillosis, Calf scour, Calf diarrhea

Escherichia coli are present in large numbers in the normal intestinal flora of humans and animals, where it generally causes no harm. However, in other parts of the body, *E. coli* can cause serious disease such as urinary tract infections, bacteraemia and meningitis. Some strains can cause severe foodborne disease. *E.*

coli infection is usually transmitted through consumption of contaminated water or food, such as undercooked meat products and raw milk. The disease in animals and humans is characterized by enteritis, abdominal cramps and diarrhea, which may be bloody. Fever and vomiting may also occur. Most patients recover within 10 days, although in a few cases the disease may become life-threatening. *E. coli* is a predominant cause of enteric disorders in neonatal calves, piglets and infants. Adults are occasionally affected.

Etiology

The disease is caused by different strains of *E. coli*, which is a Gram-negative rod-shaped, motile and facultative anaerobe. *E. coli* strains are classified into four groups on the basis of their pathogenicity:

1. **Enteropathogenic *E. coli*:** It produces diarrhea in infants.
2. **Enterotoxigenic *E. coli*:** It produces diarrhea due to enterotoxin production.
3. **Enteroinvasive *E. coli*:** This strain invades the enterocytes and produces shigellosis like dysentery in humans. It invades the mucosa of the colon.
4. **Enterohemorrhagic *E. coli*:** The strains of this group are also called Verocytotoxic *E. coli* due their ability to cause toxicity to African green monkey kidney (Vero) cells. Enterohemorrhagic *E. coli* (EHEC) is a bacterium that can cause severe foodborne disease. Infections caused by EHEC are of zoonotic importance. Primary sources of EHEC outbreaks are raw or undercooked ground meat products, raw milk and faecal contamination of vegetables. It produces verotoxin. The important pathogen of this group is *E. coli* O157:H7. In most cases, the illness is self-limiting, but it may lead to a life-threatening disease including hemorrhagic colitis (HC), hemolytic uremic syndrome (HUS), and thrombotic thrombocytopenic perpura (TTP) especially in young children and the elderly. Its significance as a public health problem was recognized in 1982, following an outbreak in the USA.

Epidemiology

The disease occurs worldwide. Cattle are the main reservoirs of Enterohemorrhagic *E. coli* (*E. coli* O157:H7). The organism is also present in the apparently healthy animals. An increasing number of outbreaks are associated with the consumption of fruits and vegetables (sprouts, spinach, lettuce, coleslaw, salad), whereby contamination may be due to contact with faeces from domestic or wild animals at some stage during cultivation or handling. EHEC has also been isolated from water bodies and has been found to survive for months in manure and water-trough sediments. Waterborne transmission has been reported, both from contaminated drinking water and from recreational waters. It has been reported from all the states of India.

Transmission

The reservoir of this pathogen appears to be mainly cattle. In addition, other ruminants such as sheep, goats, deer are considered significant reservoirs, while other mammals such as pigs, horses, rabbits, dogs, cats and birds like chickens and turkeys have been occasionally found infected. *E. coli* O157:H7 is transmitted to humans primarily through consumption of contaminated foods such as raw or undercooked ground meat products, hamburgers, salami and raw milk and milk products like cheese and yogurt. Faecal contamination of water and other foods, as well as cross-contamination during food preparation (with beef and other meat products, contaminated surfaces and kitchen utensils), may also lead to infection. Person-to-person contact is an important mode of transmission through the oral-faecal route. An asymptomatic carrier state has been reported, where individuals show no clinical signs of disease but are capable of infecting others.

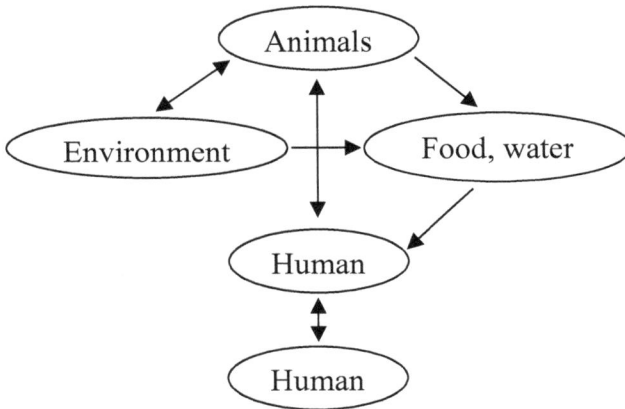

Figure 26.2: Infection Cycle of *E. coli.*

Disease in Animals

E. coli infection in animals causes four major forms:

1. **Enterotoxigenic disease:** It is characterized by diarrhea without fever in the first week of birth.

2. **Enterotoxaemic disease:** The toxin is produced by organism in the intestine of the animal which acts on the body systems. It causes enteritis in newborn and young animals.

3. **Septicemia:** It is characterized by bacteremia and extra-intestinal localization of the organisms.

4. **Invasive form:** Local invasive *E. coli* causes severe damage and ulceration in the intestine.

Disease in Humans

E. coli causes acute and chronic enteritis in human beings. The enteropathogenic *E. coli* causes vomiting and diarrhea. The Enterotoxigenic *E. coli* produces rice-water like diarrhea which resembles the cholera. Symptoms of the diseases caused by EHEC include abdominal cramps and diarrhea that may in some cases progress to bloody diarrhea (haemorrhagic colitis). Fever and vomiting may also occur. The incubation period can range from three to eight days. Most patients recover within 10 days, but in a small proportion of patients (particularly young children and the elderly), the infection may lead to a life-threatening disease, such as haemolytic uremic syndrome (HUS). HUS is characterized by acute renal failure, hemolytic anemia and thrombocytopenia. It is estimated that up to 10 per cent of patients with EHEC infection may develop HUS, with a case-fatality rate ranging from 3 to 5 per cent. Overall, HUS is the most common cause of acute renal failure in young children. It can cause neurological complications such as seizure, stroke and coma in 25 per cent of HUS patients.

Diagnosis

1. **Clinical examination:** Enteric infection usually occurs in young with watery diarrhea.

2. **Cultural examination:** The organism can be isolated and identified from clinical materials and contaminated foods. MacConkey agar and Eosin methylene blue (EMB) agar are the media of choice for isolation of *E. coli*. The colonies on MacConkey agar appear small, round and pink colour. EMB is used for the isolation and differentiation of Gram-negative enterobacteria from clinical and non clinical specimens. The colonies of *E. coli* show metallic sheen on EMB agar. MUG (4-methylumbelliferyl-ß-D-glucuronide) sorbitol agar is used for isolation of *E. coli* O157:H7 strain.

3. **Serological examination:** ELISA can be used for detection of organisms in the food. Agglutination test is used for the assay of the toxins.

4. **Molecular examination:** PCR&DNA probe can be used for detection of *E. coli*.

Disease Management in Animals

1. **Vaccination:** The vaccine containing *E. coli* K99 antigen can be used to protect the pregnant cattle and ewes. Pregnant sows can be protected by using vaccine containing *E. coli* K88 antigen.

2. **Treatment:** Chloramphenicol (@ of 20-25 mg/kg body weight, daily for one week) is recommended for treatment of the disease. Furazolidone (@ of 60 mg/Kg body weight for one week) can be used as an alternative drug.

3. **Hygienic measures:** Good hygienic standards must be adopted for effective prevention and control of colibacillosis on a farm.

Disease Management in Humans

1. **Hygienic measures:** The prevention of infection requires control measures at all stages of the food chain, from agricultural production on the farm to processing, manufacturing and preparation of foods in both commercial establishments and household kitchens. Good hygienic slaughtering practices reduce contamination of carcases by faeces, but do not guarantee the absence of EHEC from products. Preventive measures

for *E. coli* O157:H7 infection is similar to those recommended for other foodborne diseases. Ensure the proper washing of fruits and vegetables, if consumed raw. If possible, vegetables and fruits should be peeled. Regular hand washing, particularly before food preparation or consumption and after toilet contact, is highly recommended, especially for people who take care of small children, the elderly or immunocompromised individuals.

2. **Prevention of contamination of food and water:** A number of EHEC infections have been caused by contact with recreational water. Therefore, it is also important to protect such water areas, as well as drinking-water sources, from animal waste.

3. **Heat treatment of food:** The only effective method of eliminating EHEC from foods is adequate heat treatment or irradiation.

4. **Treatment:** Persons who experience bloody diarrhea or severe abdominal cramps should seek medical care. Antibiotics are not part of the treatment of patients with EHEC disease and may possibly increase the risk of subsequent HUS.

5. **Health education:** Education in hygienic handling of foods for workers at farms, abattoirs and those involved in the food production is essential to keep microbiological contamination to a minimum.

6. **WHO response:** WHO response is important in controlling foodborne diseases worldwide. During *E. coli* outbreaks, such as the ones in Europe in 2011, WHO has responded by supporting the coordination of information sharing and collaboration through International Health Regulations (IHR) and the International Food Safety Authorities Network (INFOSAN) worldwide; working closely with national health authorities and international partners, providing technical assistance and the latest information on the outbreak. In terms of prevention, WHO has responded with a global strategy to decrease the burden of foodborne diseases.

Salmonellosis

The term salmonellosis covers a complex group of infections affecting both humans and animals. The term food poisoning is also commonly applied to salmonellosis. It is one of the most common and widely distributed foodborne diseases. It constitutes a major public health burden and represents a significant cost in many countries. It is estimated that tens of millions of human cases occur worldwide every year and the disease results in more than 100 000 deaths.

Etiology

The causative agent is *Salmonella* spp., which is a Gram-negative rod-shaped bacterium. It is relatively resistant to various environmental factors. At present, more than 2579 *Salmonella* serovars are known and cause salmonellosis. The most common non-typhoidal serovar isolated from human is *S.* Typhimurium followed by *S.* Enteritidis, whereas, the most common serovar of non-human origin is *S.* Typhimurium followed by *S.* Newport. Among all the serovars of *Salmonella enterica*,

Salmonella Typhimurium is most commonly associated with enteric infections in man and animals. This serovar has diverse host range, which includes humans, cattle, pigs, sheep, horses, rodents and birds.

Epidemiology

Salmonellosis is one of the most common and widely distributed foodborne diseases. It constitute major public health burden and represents a significant cost in many countries. Millions of human cases are reported worldwide every year and the disease results in thousands of deaths. Salmonellosis occurs sporadically or in small outbreaks mostly from contaminated food and water. Animal owners, butchers, sewage and canal workers and personnel involved in the animal food processing are at risk of infection. *Salmonella* organisms have been isolated from humans and various animal species like cattle, buffalo, sheep, goat, horse, pig, poultry, camel, dog, cat and wild animals. The main reservoirs of *Salmonella* are poultry and pigs.

Transmission

1. **Ingestion:** This is the most common method of transmission. The disease is transmitted mainly through ingestion of contaminated food or water. Transmission of disease depends on food handling methods, food processing, food habits, storage, distribution methods, customs and prevailing sanitary conditions.

2. **Contact:** Contact infection may occur due to direct contact with infected animals such as dogs, pigeons, rats, mice, insects *etc.* Infection can also be transmitted by indirect contact with contaminated foodstuffs such as poultry meat, hamburger meat *etc.* Infected person can serve as chronic carrier which can serve as a source of infection for many years.

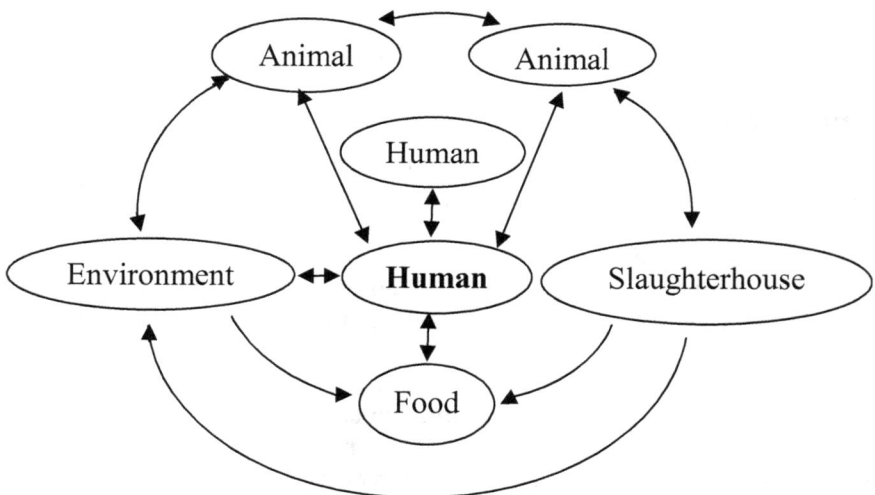

Figure 26.3: Infection Cycle of Salmonellosis.

Disease in Animals

The disease in animals is found in the form of enteritis, gastroenteritis and paratyphoid infections. Generally, the infected adult animal shows septicemia, dysentery and abortion. In calves the disease is characterized by fever, bloody diarrhea, lethargy and swollen joints with high mortality.

Disease in Humans

The incubation period varies from 6 to 72 hours. The disease is characterized by sudden onset of nausea, vomiting, foul-smelling diarrhea and fever. The stool may contain mucous and blood. Complications may lead to systemic infections and cause arthritis, osteomyelitis, pyelonephritis, peritonitis, cholecystitis, meningitis and endocarditis.

Diagnosis

1. **Cultural examination:** Confirmatory diagnosis can be made by isolation of the organism from stool or vomit of humans and tissue specimens from the intestine or other organs of animals. The selective media like brilliant green agar (BGA), bismuth sulphite agar (BSA), xylose lysine deoxycholate (XLD) agar and Hektoen enteric agar (HEA) are useful for isolation of *Salmonella* spp. Brilliant green agar is an excellent selective as well as differential medium for *Salmonella* except *S.* Typhi. Brilliant green dye present in the medium inhibits the growth of *E. coli, Proteus and* other commensal enterobacteria. On BGA, *Salmonella* produces pink colour colonies surrounded by red medium. On BSA, XLD and HEA, *Salmonella* produces black colour colonies.

2. **Serological examination:** The Widal test is valuable only for the diagnosis of typhoid and paratyphoid fever.

Disease Management in Animals

1. **Hygienic and other measures:** Hygienic standards must be maintained on the farms. Hygienic feed should be provided to the animals. Overcrowding of the animals should be avoided.

2. **Vaccination:** Animals and birds should be vaccinated with killed cultures (bacterins) or attenuated strains.

3. **Treatment:** Salmonellosis can be treated by using antibiotics such as chloramphenicol, neomycin or other drugs. Carrier animals are not treated with antibiotics but vaccinated with an attenuated *Salmonella* vaccine.

Disease Management in Humans

1. **Hygienic measures:** Purification and protection of drinking water supply, improvement of basic sanitation and promotion of food hygiene are important measures to prevent the transmission of disease. Patients and carriers should not be allowed to work in any food processing and food serving establishments.

2. **Treatment:** Generally antibiotic treatment is not required in case of acute *Salmonella* gastroenteritis. Antibiotic therapy is indicated in severe cases, infants, small children and immuno-deficient patients. Ciprofloxacin is now the drug of choice, and can be used @ 500 mg orally, twice a day for 10 days. Amoxycillin @ 0.5-1.0 g orally, four times a day for one week can be used as an alternative antibiotic for the treatment of salmonellosis.

3. **Heat treatment of food:** Poultry meat, eggs, pork *etc.* should be thoroughly cooked before their consumption.

4. **Other measures:** Salmonella testing should be carried out in poultry breeding farms. Control of rodents and prevention of wild birds from mingling with poultry on farms are also important measures to prevent the spread of the disease.

Staphylococcal Food Intoxication

Staphylococcal food-poisoning are reported worldwide in the animal and humans. A number of food and food products are associated with the outbreak of staphylococcal food-poisoning in human communities. Bovine and ovine milk, meat, poultry sausages and fish are the major sources of staphylococci.

Etiology

Staphylococcal food intoxication is caused by enterotoxins of certain strains of coagulase positive *Staphylococcus aureus*. The organism is Gram-positive, spherical cells, non-motile and non-spore forming. There are six known enterotoxins (A, B, C1, C2, D, E and F) produced by *S. aureus.* Toxins of *S. aureus* are heat-stable and can remain in the food after the organisms have been died. Toxins directly act on the intestine and central nervous system.

Epidemiology

Staphylococcal food intoxication has been reported worldwide in animals and humans. Enterotoxigenic *S. aureus* is prevalent in sheep, goat and chicken.

Transmission

Staphylococcal food intoxication results from ingestion of toxins pre-formed in the food in which bacteria have been grown (intradietetic).

Disease in Humans

The disease in humans is characterized by sudden onset of vomiting, abdominal cramps and diarrhea. In severe cases, blood and mucous may appear. Unlike *Salmonella* food poisoning, staphylococcal food intoxication rarely causes fever.

Diagnosis

1. **Clinical examination:** Food poisoning symptoms are manifested after a short period of consumption of contaminated food (as the toxins are preformed in the food).

2. **Cultural examination:** The organism can be isolated from food, vomit and faeces. Selective media like mannitol salt agar (MSA), *Staphylococcus aureus* medium No. 110 (Oxoid) and Baird-Parker agar (egg-yolk glycine tellurite pyruvate agar) can be used for isolation of staphylococci.

Disease Management in Humans

1. **Prevention of contamination of food:** Prevention of contamination of food is the best strategy to keep the food safe from staphylococcal food intoxication, because once the food is contaminated with this organism; it cannot be rendered safe event with the heat treatment, as the organism produces thermo-stable toxins. Practically it is impossible to keep all foods free from *S. aureus* due to ubiquitous nature of the organism. However, some preventive measures can be taken to render the food safe for consumption. People that are sick should not be allowed to handle the food.

2. **Prevention of growth and toxin production:** The main control of *S. aureus* is to hold the food at a temperature unsatisfactory for growth. There is little or no growth bellow 4°C or above 46°C. In fermented products, *S. aureus* can be controlled by lowering the pH with starter cultures or chemical acidulation. A combination of acidulation and starter cultures is more effective in inhibiting *S. aureus* than either treatment alone. The pH range of toxin production is about 5 to 9.

3. **Hygienic measures:** Strict hygienic conditions must be maintained during all the stages of production, processing, storage, distribution and preparation of foods to avoid contamination.

4. **Heat treatment of food:** Foods like milk, meat, poultry sausages, fish *etc.* should be adequately heat treated to kill the organisms.

Vibrioses

Vibrioses include different infectious diseases such as cholera, *Vibrio parahaemolyticus* infection, gastroenteritis, wound infection, cellulitis, sepsis *etc.*

Cholera

Cholera is an acute diarrheal infection caused by ingestion of food or water contaminated with the bacterium *Vibrio cholerae*. During the 19th century, cholera spread across the world from its original reservoir in the Ganges delta in India. Six subsequent pandemics killed millions of people across the world. The seventh pandemic started in South Asia in 1961, and reached Africa in 1971 and the Americas in 1991. Cholera is now endemic in many countries.

Etiology

The disease is caused by *Vibrio cholerae* strains. Two serogroups of *V. cholerae* (O1 and O139) cause disease outbreaks. *V. cholerae* O1 causes the majority of outbreaks, while O139 first identified in Bangladesh in 1992, which is confined to Southeast

Asia. Non-O1 and non-O139 *V. cholerae* can cause mild diarrhea but do not cause epidemics. Recently, new variant strains have been detected in several parts of Asia and Africa. Observations suggest that these strains cause more severe cholera with higher case fatality rates.

Epidemiology

Cholera may occur in epidemic and pandemic forms. Cholera is prevalent mainly in Asia, Europe and Africa. Several epidemics of cholera originated from India and spread to Western countries. Epidemics of cholera are frequent and affect children as well as adults. The consequences of a humanitarian crisis such as disruption of water and sanitation systems, or the displacement of populations to inadequate and overcrowded camps can increase the risk of cholera transmission. Dead bodies have never been reported as the source of epidemics. The short incubation period of 2 hours to 5 days, is one factor that triggers the potentially explosive pattern of outbreaks.

Transmission

The main reservoirs of *V. cholerae* are people and aquatic sources such as brackish water and estuaries, often associated with algal blooms. Recent studies indicate that global warming creates a favourable environment for the bacteria. Cholera transmission is closely linked to inadequate environmental management. Typical at-risk areas include periurban slums, where basic infrastructure is not available, as well as camps for internally displaced persons or refugees, where minimum requirements of clean water and sanitation are not met.

Disease in Humans

Cholera is an extremely virulent disease. It affects both children and adults and can kill within hours. About 80 per cent of people infected with *V. cholerae* do not develop any symptoms, although the bacteria are present in their faeces for 1-10 days after infection and are shed back into the environment, potentially infecting other people. Among people who develop symptoms, 80 per cent have mild or moderate symptoms, while around 20 per cent develop acute watery diarrhea with severe dehydration. Untreated disease can lead to death of persons.

Diagnosis

1. **Clinical examination:** The disease can be diagnosed on the basis of clinical manifestations.
2. **Cultural examination:** Thiosulphate-Citrate Bile salt Sucrose (TCBS) agar is recommended for selective isolation and cultivation of *Vibrio cholerae* and other enteropathogenic *Vibrios* causing food poisoning. Strains of *V. cholerae* produce yellow colonies on TCBS agar because of fermentation of sucrose. A few strains of *V. cholerae* may appear green or colourless on TCBS agar due to delayed sucrose fermentation.

Disease Management in Humans

A multidisciplinary approach is key for reducing cholera outbreaks, controlling cholera in endemic areas and reducing deaths. Some important approaches are discussed as follows:

1. **Hygienic measures:** Water and sanitation interventions are the long-term solution for cholera control lies in economic development and universal access to safe drinking water and adequate sanitation, which is key in preventing both epidemic and endemic cholera. Actions targeting environmental conditions include the development of piped water systems with water treatment facilities (chlorination); interventions at the household level (water filtration, water chemical or solar disinfection, safe water storage containers); and as well as the construction of systems for sewage disposal.

2. **Health education:** Health education campaigns, adapted to local culture and beliefs, should promote the adoption of appropriate hygiene practices such as hand-washing with soap, safe preparation and storage of food and breastfeeding. Awareness campaigns during outbreaks also encourage people with symptoms to seek immediate health care.

3. **Vaccination:** Currently there are 2 WHO pre-qualified oral cholera vaccines (OCVs) that are Dukoral and Shanchol. Dukoral is administered to adults and children aged more than 6 years in 2 doses; and to children aged more than 2 years and less than 6 years in 3 doses. Booster dose is administered after 6 months. Protection can be expected one week after the last dose. This vaccine is not licensed for use in children aged less than 2 years. Shanchol is given in 2 doses at an interval of 2 weeks in persons more than 1 year of age. Shanchol has provided longer term protection than Dukoral in children aged less than 5 years, and therefore, booster dose after 6 months in this age group is not required.

4. **Treatment:** Cholera is an easily treatable disease. Up to 80 per cent of people can be treated successfully through prompt administration of oral rehydration salts (WHO/UNICEF ORS standard sachet). Very severely dehydrated patients require the administration of intravenous fluids. These patients also need appropriate antibiotics to diminish the duration of diarrhea, reduce the volume of rehydration fluids needed, and shorten the duration of *V. cholerae* excretion. Mass administration of antibiotics is not recommended because it has no effect on the spread of cholera and contributes to increasing antimicrobial resistance.

5. **WHO response:** Through the WHO Global Task Force on Cholera Control, WHO works to support the design and implementation of global strategies to contribute to capacity development for cholera prevention and control globally. The WHO provides a forum for technical exchange, coordination, and cooperation on cholera-related activities to strengthen the capacity of countries for prevention and control of cholera. WHO also supports countries for the implementation of effective cholera control strategies

and monitoring of progress. It also disseminates the technical guidelines and operational manuals and supports the development of a research agenda with emphasis on evaluating innovative approaches to cholera prevention and control in affected countries.

6. **Surveillance:** Under the International Health Regulations (IHR), notification of all cases of cholera is no longer mandatory. However, cholera surveillance should be part of an integrated disease surveillance system that includes feedback at the local level and information-sharing at the global level. Countries neighboring cholera-affected areas are encouraged to strengthen disease surveillance and national preparedness to rapidly detect and respond to outbreaks of cholera spread across borders. Further, information should be provided to travellers and the community on the potential risks and symptoms of cholera, together with precautions to avoid cholera, and when and where to report cases.

Vibrio parahaemolyticus Infection

Etiology

Vibrio parahaemolyticus is a halophilic, Gram-negative, motile, straight or curved rod. Strains producing heat-stable haemolysin are considered virulent. In order to determine pathogenicity, *V. parahaemolyticus* strains are tested for positive Kanagawa phenomenon. The phenomenon is based on the ability of virulent strains to produce Kangawa haemolysin on Wagatsuma agar containing human blood. Kangawa positive strains also yield a positive ileal loop test in rabbits and vascular permeability reaction.

Epidemiology

The organism occurs throughout the world in the marine environment and coastal water; especially those containing high amount of organic matter. The microbial load is usually high inshore and estuarine areas where ambient temperatures rise seasonally and facilitate the growth of *V. parahaemolyticus*. The organism is isolated readily from coastal water, sediment, fish, shellfish, crustaceans, molluscs and faeces of the patients. In India, the organism has been isolated from sea foods and faeces of patients. The organism has also been recovered from the coastal water in Southeast Asia. Gastroenteritis caused by *V. parahaemolyticus* constitutes about 45-70 per cent of all foodborne disease outbreak in Japan.

Transmission

Vibrios are usually transmitted through ingestion of contaminated seafood.

Disease in Humans

The symptoms of illness appear within 9-25 hours of the consumption of contaminated food and lasts for 2-3 days. The disease is clinically manifested by watery diarrhea, abdominal cramps, nausea, vomiting, headache, fever and chills.

Diagnosis

1. **Clinical examination:** The disease can be diagnosed on the basis of clinical manifestations.

2. **Cultural examination:** Thiosulphate-Citrate Bile salt Sucrose (TCBS) agar is a selective medium for isolation *V. parahaemolyticus. V. parahaemolyticus* is a sucrose non-fermenting organism, and therefore produces blue-green colonies on the medium.

Disease Management in Humans

1. **Hygienic measures:** The production, handling and processing of seafood should be under hygienic conditions.

2. **Heat treatment of seafood:** Consumption of raw seafood should be avoided. The prompt heat treatment of foods destroys the viable cells and reduces the opportunities for the production of thermostable haemolysin.

3. **Prevention of recontamination of foods:** Heat treated foods should be prevented from recontamination.

4. **Treatment:** Generally, antibiotic treatment is not required as the disease is self-limited. Only symptomatic treatment is required to manage the condition of the patients.

Yersinioses

Yersinioses are group of infections, caused by *Yersinia enterocolitica* and *Y. pseudotuberculosis*. It is prevalent in humans, animals and environmental sources.

Etiology

Yersinioses are caused by *Yersinia* spp. which is Gram-negative rods belongs to the family Enterobacteriaceae. It is widespread in the intestinal tract of animals and is readily recovered from the environment, including water and soil. It can grow at refrigeration temperature.

Epidemiology

Yersinioses occur worldwide with increasing frequency. There are more than 50 serovars of *Y. enterocolitica*. Serogroups O:3 and O:9 are of significance in human infections in Europe, while serogroup O:8 is a major serovar affecting humans in the United States. O:5 and O:27 are the major serovars affecting humans in Japan and Canada. There are 14 serovars of *Y. pseudotuberculosis*, among which serogroup O:1 causes approximately 70 per cent followed by O:2 and O:3 cause 20 per cent of human infection in Europe. Serogroups O:4, O:5 and O:6 occur more frequently in Japan and less frequently in Europe.

Y. enterocolitica has been found in humans, warm blooded animals (particularly in farm animals and pets), birds and rarely in reptiles, fish and shellfish. It has been found in the intestinal tracts of pigs, dogs and cats without any clinical manifestations. Pig is very important reservoir of *Y. enterocolitica*. Strains of *Y.*

enterocolitica pathogenic for humans are found in the intestinal tract of pig, dog and cat and also in animal foodstuffs. *Y. enterocolitica* infections in humans occur more frequently during late fall and winter, while *Y. pseudotuberculosis* infections show peak in the late fall, winter and spring. *Y. pseudotuberculosis* causes infections particularly in birds, rabbits and guinea pigs. Butchers and immuno-suppressed persons are at risk of acquiring infections.

Transmission

The disease is transmitted by consumption of contaminated foods like pork and milk and milk products. Contact infection may also occur.

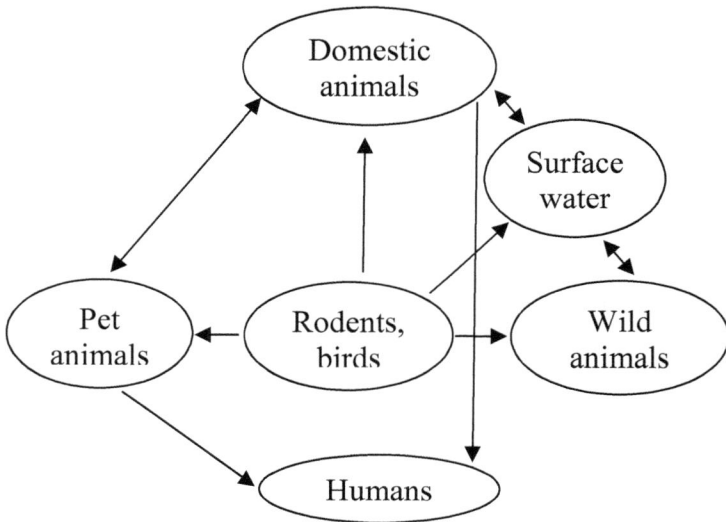

Figure 26.4: Infection Cycle of Yersinioses.

Disease in Animals

The infections in animals caused by *Y. enterocolitica* are mostly inapparent. It causes enteritis and septicemia in the animals with involvement of some other organs. *Y. pseudotuberculosis* causes acute and subacute illness. It causes granulomatous disease of internal organs in rodents.

Disease in Humans

The incubation period for *Y. enterocolitica* may range from 3 to 10 days, while for *Y. pseudotuberculosis* may be 7-21 days. Yersiniae frequently cause enteritis, enterocolitis and septicemia in immunocompromised persons. *Y. enterocolitica* most frequently causes acute ileitis, while *Y. pseudotuberculosis* most frequently causes mesenterial lymphadenitis. Extra-intestinal infections caused by *Y. enterocolitica* are meningitis, myocarditis, cholengitis, peritonitis, conjunctivitis, pneumonia, pulmonary abscess and anemia. *Y. pseudotuberculosis* may also cause extra-intestinal infections like conjunctivitis and nephritis.

Diagnosis

1. **Cultural examination:** *Yersinia* spp. can be isolated from diarrheic stool specimen by culture on cefsulodin-Irgasan-novobiocin agar. It forms red bulls eye colonies on this medium. On blood agar, it forms non-haemolytic, smooth translucent colonies, 2-3 mm in diameter in 48 hours. On MacConkey agar, it forms pinpoint colourless colonies at 22°C. Specimens from mesenteric lymph nodes, tissue samples obtained during surgery, pus and blood from septicemic cases can also be used for isolation of the organism. Yersinioses in animals can be diagnosed by isolation of organisms from faeces and organs.

2. **Serological examination:** The agglutination reaction can be used for detection of antibodies against *Yersinia* spp. ELISA is also useful for diagnosis of *Yersinia* infections.

3. **Molecular examination:** PCR has been useful for detection of organism in clinical and food specimens.

Disease Management in Animals

The disease in the animals can be prevented by strict hygienic measures. There should be proper disposal of excretions and other wastes. Oxytetracycline is effective in the treatment of infection in animals.

Disease Management in Humans

1. **Hygienic measures:** Food hygiene is important for prevention of yersinioses. Contact with infected animals should be avoided for prevention of yersinioses. The food industry plays a critical role in preventing yersiniosis. It is important to reduce the level of contamination of raw pork products with *Yersinia* by improving the hygiene at slaughter. Contamination of dairy products can be prevented by maintaining high standards of hygiene in dairy plants.

2. **Chemotherapy:** Treatment is recommended only in severe infections like septicemic, chronic and recurrent enteric yersinioses. Ciprofloxacin @ 500 mg twice a day or doxycycline @ 100 mg twice a day, orally for one week are effective.

3. **Heat treatment of milk, meat and their products:** Adequate heat treatment of milk, meat their products can reduce the spread of the disease in humans.

Section - III

Viral Zoonoses

Viral zoonoses are one of the important zoonoses causing potential public health hazards. Some of them cause fatal diseases in animals and humans. Viral zoonoses can be grouped on the basis of type of infection they produce in natural host. These include viral zoonoses causing encephalitis, hemorrhages and local lesions. Viral zoonoses causing encephalitis belong mostly to Rhabdoviridae, Flaviviridae, Togaviridae, Reoviridae and Bunyaviridae. Most of them are transmitted through mosquito or tick bites, except a few which are transmitted through bite of an infected host, *e.g.*, rabies. Most of the viral zoonoses causing hemorrhagic fevers are reported to be of emerging and re-emerging in nature. These viruses belong mainly to Arenaviridae, Bunyaviridae, Flaviviridae and Filoviridae. These infections are often associated with extensive hemorrhages in humans. Most of them are transmitted through vectors (mosquitoes and ticks). Some viral zoonoses are associated with local rashes and arthralgia. Most of these viruses belong to the family Togaviridae. Most of them are transmitted to humans through infected mosquito bites. Other zoonosis like prion diseases are caused by scrapie associated prion proteins (PrPsc), which are proteinaceous infectious agents common in animals and humans. The human disease variant (vCJD) is believed to be a zoonotic disease caused by BSE agent and transmitted through exposure to food contaminated by the bovine BSE agent. Emerging and re-emerging diseases cause devastating effects internationally and affects millions of people. The major factors that contribute to emergence and re-emergence of diseases are environmental (ecological) changes, microbial adaptation and change, technological changes, farming changes and international travel and trade. Moreover, movement of population, birds, vectors, pathogens and trade contribute to the global spread of emerging infectious diseases such as influenza, severe acute respiratory syndrome (SARS) *etc*. Other factors *viz*., human migration, change in land use pattern, mining, coastal land degradation, wetland modification, construction of buildings, habitat fragmentation, deforestation, expansion of agents host range, human intervention in wildlife resources like hiking, camping and hunting also influence on acquiring zoonotic infections from wildlife. The RNA viruses are capable of adapting to changing environmental conditions rapidly and are among the most prominent emerging pathogens. Mutations are more common in RNA viruses than DNA viruses. Some important viral zoonoses are discussed in next chapters.

Chapter 27

Zoonoses Caused by Rhabdoviruses

Rhabdoviruses belong to the family Rhabdoviridae. The hosts of viruses constituting the family Rhabdoviridae include mammals, birds, reptiles, fish, insects and plants. The members of the family are enveloped, single-stranded RNA viruses, with bullet-shaped or rod-shaped morphology. The family includes three genera namely, *Lyssavirus*, *Vesiculovirus* and *Ephemerovirus*. *Lyssavirus* and *Vesiculovirus* are of zoonotic importance. The name *Lyssavirus* has been derived from the Greek word for madness or frenzy. The members of the genus *Vesiculovirus* are associated with the disease like vesicular stomatitis, is particularly prevalent among cattle, horses and pigs.

Rabies

Synonyms

Hydrophobia, Lyssa, Tollwut

Rabies is an acute, highly fatal viral disease of central nervous system. It is primarily a zoonotic disease of warm-blooded animals, particularly carnivores such as dog, cat, jackal, wolves and mongoose.

Etiology

The causative agent is *Lyssavirus* type 1. It is bullet-shaped neurotropic RNA containing virus belong to the family Rhabdoviridae. Serotype 2, 3 and 4 are rabies-related viruses. They cause rabies like disease in animals and humans. Anti-rabies vaccines are not effective against the rabies related viruses. Dog, fox and jackal are the main reservoirs of rabies virus. The virus is excreted in the saliva of infected animals. The main source of infection in human is the saliva of rabid animals.

Epidemiology

Rabies is enzootic/endemic and epizootic disease of worldwide importance. In India rabies occurs in all the states with the exception of Lakshadweep, Andaman and Nikobar islands. Geographical boundaries play important role in the transmission of rabies. Water is the most important natural barrier to rabies. Australia, New Zealand, UK, Ireland, Norway, Sweden, islands of Western Pacific, Japan and Taiwan are rabies free. Epidemiology of rabies differs from country to country. More than 95 per cent human deaths occur in Asia and Africa. Rabies occurs in the following epidemiological forms-

1. **Urban rabies:** It is predominantly found in developing countries in Asia and Africa. The urban rabies is maintained in stray dogs. Dogs, cats or other animals are infected due to coming in contact with infected sylvan or rural reservoirs.

2. **Sylvatic rabies (Wildlife rabies):** It is most commonly found in developed countries in Northern hemisphere. The sylvatic rabies is perpetuated by jackal, fox, hyena, mongoose *etc.*

3. **Bat rabies:** It belongs both to the urban and sylvan type. Transmission of rabies in animals and humans occurs through bite of vampire bats. Transmission in humans occurs when they sleep outdoors. Vampire bats have not been reported in India.

Transmission

Rabies may be transmitted through the following routes-

1. **Animal bites:** This is the most common mode of transmission of rabies. Transmission of rabies to human beings oftenly occurs through rabid dog bites. The disease may also be transmitted through bite of other infected or rabid animals.

2. **Licks:** Transmission of infection may occur through licks on abraded skin and abraded and unabraded mucosa.

3. **Aerosol transmission:** This may occur in the laboratory dealing with infected tissues especially during homogenization. This route of transmission may also occur in caves harboring rabies infected bats.

4. **Person-to-person transmission:** This mode of transmission is rare but possible in case of corneal and other organ transplants. Rabies in human is a dead-end infection.

Disease in Animals

Rabies in dogs may manifest itself in two forms-

1. **Furious rabies:** This is the typical "mad dog syndrome", characterized by change in behaviour, running amuck (tendency to run away from home and wander aimlessly and biting humans and animals who may come in its way), change in voice, excessive salivation and paralytic stage (paralysis of whole body leading to coma and death).

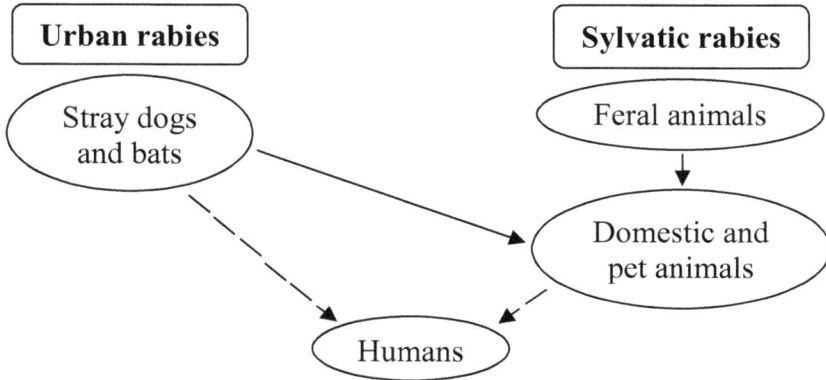

Figure 27.1: Infection Cycle of Rabies.

2. **Dumb rabies:** This is form is lacking of irritative or excitative stage. This form is predominantly paralytic. The dog withdraws itself from being seen or disturbed. It lapses into a stage of sleepiness and dies in about 3 days. Hydrophobia is absent in the animals.

Disease in Humans

The incubation period in human is highly variable, most oftenly 3-8 weeks following exposure but may vary from 4 days to many years. The incubation period depends on the site of bite, severity of bite, number of wounds, amount of virus injected and species of biting animal. In general, the incubation period is shorter in case of sever exposure and bite on face, head, neck and upper extremities and bite by wild animals. The prodromal stage is characterized by headache, nausea, vomiting, malaise, sore throat and slight fever. Oftenly, tingling pain occurs at the site of bite. This stage is followed by nervous excitation. The patient is intolerant to noise, bright light or cold air. Other changes include fear of death, anger, irritability and depression. The symptoms are progressively aggravated and all attempts at swallowing liquid become unsuccessful. The patient may die suddenly or pass on to the stage of paralysis or coma.

Diagnosis in Animals

Rabies diagnosis in the animals includes examination of negri bodies in nervous tissues, mouse inoculation test and fluorescent antibody technique.

Diagnosis in Humans

Diagnosis can be made on the basis of history of rabid animal bite and clinical symptoms. Isolation of the virus from saliva and other secretions is another method of diagnosis. Confirmatory diagnosis in the early stage of illness can be made by antigen detection using immunofluorescence of skin biopsy.

Disease Management in Animals

1. **Management of wound:** It includes cleansing of wound with soap and running water. Sodium bicarbonate or caustic soda or 2 per cent quaternary ammonium compounds or tincture iodine can be used.

2. **Anti-rabies serum:** It can be infiltrated around the wound.

3. **Vaccination:** Oral vaccine baiting is especially valuable for vaccinating foxes, mongoose and other wild animals.

4. **Treatment:** There is no specific treatment for clinical rabies. Symptomatic treatment may be rendered with sedative and narcotic drugs.

Disease Management in Humans

1. **Management of wound:** It includes cleansing of wound with soap and running water. Sodium bicarbonate or caustic soda or 2 per cent quaternary ammonium compounds or tincture iodine can be used.

2. **Anti-rabies serum:** It can be infiltrated around the wound.

3. **Vaccination:** It can be done, using human diploid cell or primary chick embryo vaccine in persons at risk such as veterinarians, animal owners and forest workers.

4. **Treatment:** There is no specific treatment for rabies. The case management includes isolation of patients in a quiet room, protection from external stimuli and maintaining fluids and electrolytes. Treatment with interferon alone or in combination with ribavirin can be attempted.

Chapter 28

Zoonoses Caused by Prions

Transmissible spongiform encephalopathies (TSE) are the group of diseases in animals and humans associated with central nervous system. TSE have very long incubation periods and are fatal diseases of animals and humans. TSE are transmitted by so-called prions, which are actually proteinaceous infectious particles. Prions apparently consist only of protein and do not have a genome themselves but are encoded by the host (animals and humans) gene, the prion protein (PrP) gene. Prions have abnormal folded structures of pleated sheets. They accumulate in intracellular vesicles in brain which increase into large vacuole and can be deposited as "florid plaques". Important transmissible spongiform encephalopathies in animals and humans are bovine spongiform encephalopathy (BSE) and Variant Creutzfeldt-Jakob disease (vCJD).

Bovine Spongiform Encephalopathy (BSE)

Synonym

Mad cow disease

Bovine spongiform encephalopathy (BSE) is a transmissible spongiform encephalopathy (TSE) of bovines. It affects a number of species including humans. BSE is a transmissible, neuro-degenerative fatal brain disease of cattle. The incubation period is very long (4-5 years). It is very fatal disease of cattle.

Etiology

The nature of the BSE agent is still being debated. Available evidence supports the theory that the agent is composed of a self-replicating protein, referred to as a prion.

Epidemiology

The disease was first reported in 1989 in the UK. Since then some cases have also appeared in other countries like Israel and Japan. BSE has also been reported in 1997, 2001 and 2002 from the UK. Cattle in the UK are continuously monitored for BSE. The number of cases of BSE has been decreasing since 1992.

Transmission

BSE is transmitted through consumption of meat and bone meal supplements in cattle feed contaminated with BSE agent. Initially, BSE outbreaks occurred in the UK, shortly afterwards, it introduced a ban on the feeding of protein derived from ruminants (*e.g.* cattle, sheep and goats) to any ruminant. The use of bovine offal in the human food chain considered to pose a potential risk, and was also banned in the UK in 1989.

Disease in Animals

Animals such as cattle, sheep, goat and cats can be affected by TSEs. The best known form of disease in animals is BSE or "Mad cow disease". The disease is characterized by progressive degenerative changes in the brain. BSE caused heavy economic loss to beef industry in the UK.

Diagnosis

The disease in animals can be diagnosed by histopathological examination of the brain (antigen detection). Antibodies against prions cannot be found in infected animals.

Disease Management in Animals

Strict hygienic measures must be followed in case of BSE outbreaks and immediately notified to the higher health authorities in order to prevent the spread of the disease. There should be ban on meat and bone meal in animal feed during BSE outbreaks. Testing of slaughtered animals, systemic removal of "high-risk material" from carcasses and destruction of suspect and confirmed bovine cases and control of animals potentially exposed at the same time are also important. Imported feeds derived from ruminants should be certified and free from BSE. Ruminant derived feeds should be sterilized before their consumption.

Variant Creutzfeldt-Jakob Disease (vCJD)

Variant Creutzfeldt-Jakob disease is a rare and fatal human neurodegenerative disease. In humans, the most common transmissible spongiform encephalopathy (TSE) is called Creutzfeldt-Jakob disease (CJD). In humans, the disease is characterized by loss of memory and ability to think and to move properly. It causes progressive damage of brain which results in the loss of vision, inability to speak and feed.

Etiology

The human disease "variant Creutzfeldt-Jakob disease" (vCJD) is believed to be a zoonotic disease caused by the BSE agent. The consumption of food of bovine origin contaminated with the BSE agent, a disease of cattle, has been strongly linked to the occurrence of vCJD in humans. There is strong scientific evidence that vCJD is linked with exposure to a BSE of cattle. The link between vCJD and BSE was first hypothesized because of the association of these two TSEs in time and place. In addition, laboratory evidence indicates that vCJD is linked causally with BSE. The most likely cause of vCJD is exposure to the BSE agent through consumption of food from bovine origin most possibly contaminated by infected bovine brain or other central nervous system tissue.

Epidemiology

Variant CJD was first recognized in March 1996. It is rare with a worldwide incidence of 1 case per million. The disease has been reported from UK, France, Canada, Italy, Portugal, Ireland, USA, Japan, Saudi Arabia and Taiwan. 175 cases of vCJD were reported in UK of Great Britain and Northern Ireland (UK) and 49 cases in other countries from October 1996 to March 2011. Following the successful containment of the BSE epidemic in cattle, the number of cases of vCJD in the UK has declined since 2000.

Transmission

The route of transmission of vCJD is not yet fully proven but it is generally accepted that it is transmitted through exposure to food contaminated by the bovine BSE agent. The cases of vCJD infection have also been associated with blood transfusion.

Disease in Humans

Early in the illness, patients usually experience psychiatric or sensory symptoms, which most commonly take the form of depression, apathy or anxiety and occasionally unusual persistent and painful sensory symptoms. Neurological signs such as unsteadiness, difficulty in walking and involuntary movements, develop as the illness progresses. By the time of death, patients become completely immobile and mute. The disease exists in three forms-

1. **Sporadic CJD:** It occurs throughout the world at the rate of about one per million people, and accounts for about 85 per cent of CJD cases.

2. **Familial CJD:** It occurs due to mutation of a gene and accounts for about 5-15 per cent of CJD cases.

3. **Iatrogenic CJD:** It results from accidental transmission via contaminated surgical equipment or as a result of corneal or meningeal transplants or administration of human-derived pituitary growth hormones. It accounts for less than 5 per cent CJD cases.

Diagnosis

The disease is clinically characterized by progressive nature of the disease and failure to find any other diagnosis.

There are no completely reliable tests to use before the onset of clinical symptoms. However, magnetic resonance scans and tonsillar biopsy are useful diagnostic tests. Currently, the diagnosis of vCJD can only be confirmed following pathological examination of the brain post-mortem. Characteristically, multiple microscopic and abnormal aggregates encircled by holes are seen in the brain tissue, resulting in a daisy-like appearance described by the term "florid plaques".

Disease Management in Humans

1. **Prevention of contamination of food:** BSE agent containing animal tissues must be prevented to enter the animal or human food chain.

2. **Ban on ruminant derived food:** All countries where BSE is known to exist should ban the use of ruminant tissues in ruminant feed. In pharmaceutical industry, the use of bovine materials and materials derived from other animal species in which TSEs naturally occur should be prohibited. However, if their use is absolutely necessary, these materials should be obtained from BSE free countries.

Chapter 29

Zoonoses Caused by Orthomyxoviruses

Orthomyxoviruses belong to the family Orthomyxoviridae. The family Orthomyxoviridae includes the influenza viruses. Influenza viruses are pathogenic to the animals and humans. The most commonly infected animals are swine, equine and birds. One of the most prominent features of the influenza viruses is their ability to change antigenically either gradually over years (antigenic drift) or suddenly (antigenic shift). On the basis of nucleoprotein and Matrix (M) capsid protein, influenza virus is of three types, *viz.*, Type A, Type B and Type C. There are two types of protein that is haemagglutinin (H) and neuraminidase (N). Haemagglutinin is of seventeen types (1-17), while neuraminidase of ten types (1-10). On the basis of combination of haemagglutinin and neuraminidase proteins, the Type A influenza virus has been categorized into several subtypes, *viz.*, H1N1, H1N2, H2N2, H3N1, H3N2, H3N8, H5N1, H5N2, H5N3, H5N8, H5N9, H7N1, H7N2, H7N3, H9N2 and H10N7. These can infect several species such as human, avian, swine, equine *etc.* Only influenza A virus has the potential to shift, whereas all three types may drift antigenically. Influenza A virus is zoonotic and B is found only in humans. Influenza C viruses are found in the animals and humans but their zoonotic role is unknown. Influenza virus Type A exhibits frequent antigenic variations. Influenza virus Type B exhibits infrequent antigenic variations. It causes epidemics and seasonal influenza. Influenza virus Type C is antigenically stable. It causes mild respiratory disease and does not cause epidemics.

Avian Influenza

Synonyms

Fowl plague, Bird flu

Avian influenza, commonly called bird flu, is an infectious viral disease of birds. It is caused by a strain of the influenza virus called "type A". Most avian influenza viruses do not infect humans but A(H5N1) and A(H7N9) causes serious infections in humans. Bird flu in human was first observed in an outbreak in Hong Kong in 1997. Influenza A virus (H5N1) caused the death of 6 of 18 infected persons. Before 1997, it was believed that only swine influenza but not avian influenza can be transmitted to humans.

Etiology

Influenza is an acute contagious disease caused by a virus belonging to family Orthomyxoviridae. The main etiological agents of avian influenza are avian influenza virus A(H5N1) and A(H7N9). Human infection with avian influenza A(H7N9) was reported for the first time, from China in February 2013. Avian Influenza virus is a RNA Orthomyxovirus. Influenza virus particle is enclosed in an envelope carrying protein (antigen) spikes, which are of two types, *viz.*, haemagglutinin (HA) and neuraminidase (NA). There 13 HA and 9 NA subtypes of influenza virus A. Changes in HA and NA composition of the virus determine the epidemiological pattern of the disease in humans and animals. Avian influenza A viruses exist in wild and domestic birds. Infections are often latent, especially in waterfowl. The virus can be transmitted among avian species and can be carried by migrating birds. A minor antigenic change especially in the HA is called antigenic drift, while, major variations in HA and NA are called antigenic shifts. Antigenic shifts are the cause of influenza pandemics, while drifts result in the occurrence of epidemics.

Epidemiology

The disease occurs worldwide. It may occur in pandemic pattern. Influenza virus (H5N1) is highly pathogenic and enzootic in many bird populations, especially in Southeast Asia. Avian influenza virus killed millions of poultry involving several countries. Wild birds, waterfowl and poultry are the main reservoirs of avian influenza virus. This virus is mainly transmitted among domestic birds through movement of infected birds, poultry products and infected poultry manure. Avian influenza is endemic in poultry in Bangladesh, China, Egypt, India, Indonesia and Viet Nam. The zoonotic importance of bird flu has been recognized in 1997. Outbreaks of bird flu have also been reported in 2002, 2006 and 2007 from many countries. Recent outbreak of bird flu has been reported in June 2008 from China, Indonesia, Vietnam, Pakistan and Egypt. In India, outbreaks of bird flu have been reported in January 2006 in Maharashtra and Gujarat and in February 2007 in Manipur. Outbreak of bird flu has also been reported in January 2008 in Birbhum district of West Bengal in which H5N1 avian influenza virus killed lacs of poultry. In India, large outbreaks of bird flu has been seen almost every year since 2010. From 2003 to 2013, of the 649 laboratory-confirmed human cases of A(H5N1) officially reported to WHO from 15 countries, 385 died. Of these, 228 cases (35 per cent) and 181 deaths (47 per cent) were from South-East Asia.

Transmission

1. **Contact:** The avian influenza virus is mainly transmitted through direct contact with infected birds. Almost all birds are susceptible to this virus.

Influenza virus (H5N1) is transmitted among birds through saliva, nasal secretions, blood and faeces of infected birds. In other animals this virus is transmitted through direct contact with bodily fluids or through contact with surfaces contaminated with infected birds. Infected birds shed the virus for long periods that is act as a reservoir. The avian influenza A virus H5N1 was transmitted to humans primarily by contact with live birds and not by contact with meat or animal products in an outbreak of avian influenza in Hong Kong in 1997. This virus may be transmitted from one person to another person.

2. **Transplacental:** The H5N1 bird flu virus may be transmitted to fetus through the placenta of infected pregnant woman.

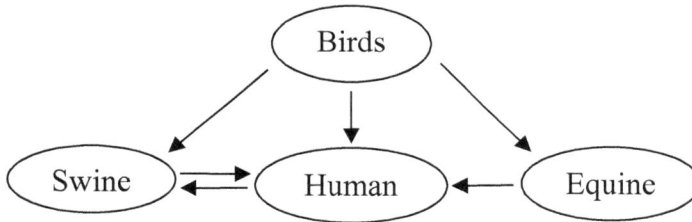

Figure 29.1: Infection Chain of Avian Influenza.

Disease in Animals

In birds, the disease is characterized by cough, sneezing, dyspnea, lacrimation, edema of head and neck, dullness, cyanosis of comb and wattle and cessation in egg production. Some birds may also show diarrhea and central nervous system involvement.

Disease in Humans

The incubation period ranges from 2-8 days, but may extend up to 17 days. The disease is characterized by onset of high fever, cough, sore throat, respiratory distress, a hoarse voice, and a crackling sound when inhaling. In some patients, the early symptoms may be diarrhea, vomiting, abdominal pain, chest pain, and bleeding from the nose and gums. The complications of the disease may lead to hypoxemia, multiple organ dysfunction, and secondary bacterial and fungal infections.

Diagnosis

The disease can be diagnosed by isolation and identification of influenza virus. This virus can be easily grown in embryonating chicken eggs or in cell culture by conventional methods. But this method is time consuming and can be suitable for conducting surveillance of influenza on worldwide basis. The PCR test is most suitable for clinical diagnosis of influenza. The differential diagnosis should be made with human influenza A virus, *Hantavirus* pulmonary syndrome (HPS) and SARS.

Disease Management in Animals

1. **Proper housing and management of poultry:** There should be provision of proper housing and ventilation in the poultry house to safeguard against rapid transmission of the virus. Overcrowding of poultry and other animals should be avoided.

2. **Quarantine:** Prevent the introduction of the virus into the poultry farm and follow up of strict quarantine is another important measure for prevention of disease.

3. **Vaccination:** Current licensed vaccines are predominately inactivated whole avian influenza vaccines. These are produced from low pathogenicity (LP) avian influenza virus strains, or occasionally from high pathogenicity (HP) avian influenza virus strains. Recently, reverse genetic procedures have been developed that allow construction of vaccine strains using a genetically altered HA gene (changing HP HA proteolytic cleavage site to LP) and a backbone of internal gene segments for safe, high growth production. Other licensed AI vaccines include recombinant fowl poxvirus vector with an AI H5 insert and a recombinant Newcastle disease virus vector with an AI H5 gene insert. The latter vaccine can be mass administered via aerosol application.

4. **Treatment:** Antibiotics may be used to check the secondary infection.

5. **Slaughter of poultry:** Slaughter of poultry and their proper and hygienic disposal are essential to restrict the spread of the disaese at poultry farm, and hence in the public.

Disease Management in Humans

1. **Vaccination:** Vaccines are available to protect from avian influenza. A new recombinant A(H5N1) vaccine virus has been developed by the National Institute of Virology, Pune, India and the WHO Collaborating Center for the Surveillance, Epidemiology and Control of Influenza at the Centers for Disease Control and Prevention, Atlanta, USA, from A/chicken/India/NIV33487/2006 (H5N1;clade 2.2) The new recombinant vaccine virus named A/India/NIV/2006(H5N1)-PR8-IBCDC-RG7 is available.

2. **Treatment:** In human beings, some antiviral drugs such as amantadine, rimantadine, oseltamivir or zanamivir can be used as prophylactically or curatively against influenza virus. Amantadine or rimantadine @ 100 mg orally, twice a day can be used. Oseltamivir (Tamiflu) @ 75 mg orally, twice a day can also be used. Zanamivir (Relenza) @ 5 mg per inhalation, twice daily can be used.

3. **Slaughter of poultry:** In ordered to prevent widespread transmission of the influenza virus, all birds of the infected poultry farm should be slaughtered without any delay.

4. **Hygienic measures:** Poultry should not be held in close contact with swine, since avian influenza virus can cross species. Slaughtered poultry must be disposed properly under strict hygiene. Litter and other soiled

materials with the infected birds must be disposed promptly with hygienic precautions. Premises inhabited by infected birds must be disinfected. There should be proper disinfection of water trough, feeder and equipment associated with poultry farm.

Swine Influenza

Swine influenza viruses can easily be transmitted from swine to humans and vice versa. A/H1N1 is a flu virus. When it was first detected in 2009, it was called "swine flu" because the virus was similar to those found in pigs. Swine flu is a respiratory disease caused by a new strain of influenza virus. It spreads rapidly from human to human, which leads to a pandemic flu outbreak. Pandemic flu is different from ordinary flu because it's a new flu virus that appears in humans and spreads very quickly from person to person worldwide. Three pandemics occurred due to influenza virus in the previous century, in 1918, 1957 and 1968. The 1918 pandemic was the most devastating, caused deaths of 30-40 million people worldwide. The subsequent pandemics were relatively milder, each killing around 1 million people. In 2009 influenza outbreak, more than 414000 cases and about 5000 deaths had been reported to WHO by 195 countries worldwide.

Etiology

Swine influenza is caused by Influenza A/H1N1 virus. The virus belongs to the family Orthomyxoviridae. Influenza A virus undergoes mutation that takes place within the genome (antigenic drift)/or reassortment among the genetic materials of subtypes (antigenic shift) resulting in a new virus. Antigenic drift is responsible for new seasonal strains that make necessary surveillance to detect these strains and to prepare new seasonal influenza vaccine (yearly basis). Antigenic shift may result in a new virus easily transmissible from human to human for which the population has no immunity that results in pandemics. Influenza A virus causes pandemics, epidemics, seasonal flu outbreaks and sporadic infections. This virus caused many pandemics as follows:

Spanish flu [A (H1N1)] : 1918-19

Asian flu [A (H2N2)] : 1957-59

Hong Kong flu [A (H3N2)] : 1968

Swine flu [A (H1N1)] : 2009-10

Epidemiology

Swine influenza seen predominantly in the mid-Western United States (and occasionally in other states), Mexico, Canada, South America, Europe (including UK, Sweden, and Italy), Kenya, Mainland China, Taiwan, Japan and other parts of Eastern Asia and in various parts of India. The A(H1N1) pandemic 2009 influenza virus contained genes from pig, bird and human influenza viruses in combination that was never reported before 2009 in any part of the world. Genetic analyses of this virus have shown that it originated from animal influenza viruses and is unrelated to the human seasonal H1N1 viruses that have been in general circulation among people since 1977. After early outbreaks in North America in April 2009, the new influenza virus spread rapidly around the world. There are three main influenza A virus subtypes (H1N1, H1N2, and H3N2) that have been isolated from pigs in the United States. H1N1 and H3N2 swine flu viruses are endemic among pig populations in the United States and something that the industry deals with routinely. Outbreaks among pigs normally occur in colder weather months (late fall and winter), but can occur year round. In 2009 influenza outbreak in India, the state of Maharashtra was the worst affected, followed by Karnataka. In Europe, the estimated attack rates were in the range of 20 per cent - 30 per cent. In Australia and New Zealand, the attack rate was 20 per cent each. In the USA and the UK, attack rates among the elderly were lower due to some exposure to the virus in the past. In India, the predominant influenza virus circulating that season was not the A(H3N2) but the A(H1N1) pandemic 2009 or swine flu.

In some cases these zoonotic infections result in severe disease or even death in humans, but often these infections result in only a mild illness or appear to cause no illness at all. All of the past four pandemic influenza viruses have contained gene components originating in animals. The actual public health risks posed by influenza viruses circulating in bird, swine, and other animal populations are not completely understood. Recent findings suggest that influenza viruses in animals and humans increasingly behave like a pool of genes circulating among multiple hosts, and that the potential exists for novel influenza viruses to be generated in swine and other animals. This situation reinforces the need for close monitoring and close collaboration between public health and veterinary authorities.

Transmission

Swine can become infected with influenza viruses from a variety of different hosts such as birds and humans, therefore, swine can act as a "mixing vessel", facilitating the reassortment of influenza genes from different viruses and creating a "new" influenza virus. The concern is that such "new" reassortant viruses may be more easily spread from person to person, or may cause more severe disease in humans than the original viruses. The virus is spread among pigs by aerosols, through direct, indirect contact, and also by asymptomatic carrier pigs. Spread among pigs mostly occurs through close contact and possibly from contaminated objects moving between infected and uninfected pigs. Infected swine herds, including those vaccinated against swine flu, may have sporadic disease, or may show only mild or no symptoms of infection. The pandemic (H1N1) 2009 virus has not been shown to be transmissible to people through eating properly handled and prepared pork or other products derived from pigs.

Swine flu is transmitted as easily as the normal seasonal flu and can be passed to other people by exposure to infected droplets expelled by coughing or sneezing that can be inhaled, or that can contaminate hands or surfaces. There is also potential for transmission through contact with fomites that are contaminated with respiratory or gastrointestinal material. There is a risk that people may also acquire infection by touching something that is contaminated by the virus and then touching their nose or mouth. Food is not yet known to be a vehicle for the transmission of this new influenza virus. Rapid spread among the population has been observed, especially in crowded places such as schools.

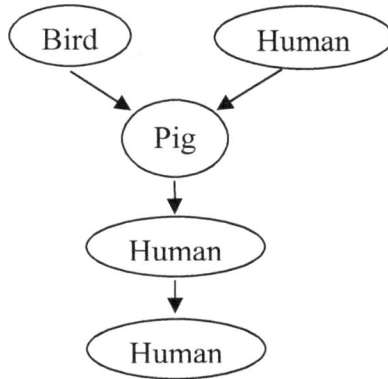

Figure 29.2: Infection Chain of Swine Influenza.

Disease in Swine

Swine influenza is a highly contagious acute respiratory disease of pigs. The disease in the pigs is characterized by high fever, nasal discharge, redness or inflammation of eyes, lethargy, sneezing, breathing difficulties, coughing (barking like sound) and anorexia. Some pigs infected with influenza, however, do not show overt disease.

Disease in Humans

The clinical spectrum of infection with the H1N1 virus is broad, and ranges from mild upper respiratory tract illness to severe complications. The majority of people with pandemic influenza experience mild illness and recover fully without treatment. Severe complications include pneumonia resulting in respiratory failure, acute respiratory distress syndrome (ARDS), multi-organ failure and death. Gastrointestinal symptoms such as diarrhea have been reported in 20 per cent –50 per cent of patients, and do not require hospitalization. Other symptoms are fever, chills, cough, sore throat, runny or stuffy nose, arthralgia, myalgia, headache, fatigue and sometimes vomiting and diarrhea.

Diagnosis

Samples for laboratory tests should be taken from the deep nasal passages, nasopharynx, throat or bronchial aspirate. Reverse transcriptase-polymerase chain reaction (RT-PCR) provides the most timely and sensitive evidence of infection.

Disease Management in Swine

1. **Proper housing and management of swine:** Overcrowding and low humidity provide ideal conditions for rapid transmission of disease. Therefore, overcrowding in the swine herds should be avoided in order to minimize the transmission of disease. Proper ventilation system in the animal houses is also helpful in the prevention of disease. Avoid contact

between swine and poultry to prevent the spread of disease, because interspecies transmission of the virus is possible.

2. **Hygienic measures:** Adopt high standards of hygiene at swine farms. Imported swine should be subjected to quarantine so that clinically inapparent swine may come in the overt disease. Good biosecurity measures should be adopted to prevent disease outbreaks.

3. **Vaccination:** A flu vaccine for pigs can help but not 100 per cent effective. Sometimes the vaccine used may not protect against viruses circulating in the pigs. In addition, current vaccines may not be effective in young pigs due to interference from antibodies received from sows.

4. **Treatment:** Symptomatic treatment can be followed. Antibiotic treatment can be attempted to check the secondary infection.

Disease Management in Humans

1. **Avoid contacts:** The virus pose a potential risks to those who are likely be in close contact with pigs, especially, pig handlers, abattoir workers and veterinarians. Therefore, contact with infected pigs should be avoided.

2. **Safety precautions:** Infected person should stay at home and rest. Avoid close contact with healthy people within the family or public place in order to avoid transmission of disease in other persons. Drink plenty of water and other clear liquids to prevent dehydration. Treatment according to the symptoms such as fever and cough can be followed. In severe infection, pregnancy and higher risk of flu complications like asthma, consultation with physician is essential.

3. **Use of mask:** The Ministry of Health and Family Welfare, Government of India has recommended the guidelines on use and the correct procedure of wearing triple layer surgical mask for healthcare workers, patients and members of public. Change the mask after six hours or as soon as they become wet. While removing mask great care must be taken not to touch potentially infected outer surface. Disposable masks are never to be reused and should be disposed of properly.

4. **Proper disposal of used masks:** Used mask is potentially infected medical waste which should be disposed of in the identified infectious waste disposal bag/container. In community settings where medical waste management protocol cannot be practiced, it may be disposed of either by burning or deep burial. Used masks should be disinfected using bleach solution or sodium hypochlorite solution or quaternary ammonium then disposed of either by burning or deep burial.

5. **Vaccination:** Annual vaccination is the most effective solution for combating seasonal influenza infections such as swine flu. It is recommended that people get a flu vaccine even during seasons when drifted viruses are circulating. It is because vaccination can prevent some infections and can reduce serious ailments that can lead to hospitalization

and death. There are two different types of flu vaccines, trivalent and quadrivalent. Trivalent vaccines protect against 3 strains of the flu (A/H3N2, A/H1N1, and influenza B). Quadrivalent vaccines protect against 4 strains of the flu (A/H3N2, A/H1N1, and 2 strains of influenza B). The WHO has recommended the following for the Northern Hemisphere's trivalent vaccines:

For H1N1, an A/Michigan/45/2015-like virus

For H3N2, an A/Hong Hong/4801/2014-like virus

For B, Brisbane/60/2008-like virus (belonging to the Victoria lineage)

For quadrivalent vaccine, the WHO has recommended adding Phuket/3073/2013-like virus, a Yamagata lineage virus that is the second B component of quadrivalent vaccines. This vaccine has been recommended for both the Southern Hemisphere's past and the Northern Hemisphere's current season.

The new H1N1 vaccine strain, called A/Michigan/45/2015, replaces A/California/7/2009, which has been in use as a vaccine strain since the 2009 H1N1 virus became a regularly circulating seasonal flu strain after the 2009-10 pandemic. WHO has recommended it for the Southern Hemisphere's 2017 vaccine to improve protection against subclades that emerged last year. The flu vaccine is available by shot or nasal spray. Swine influenza (swine flu) vaccines are commercially available with different brands such as Vaxigrip, Influgen *etc*. Vaxigrip (Influenza Vaccine (A and B), H1N1 Vaccine (Swine Flu)) is prescribed for children aged 6 months and older against influenza caused by influenza virus. Possible serious side effects of vaccination with influenza vaccine include difficulty in breathing, hoarseness, swelling around the eyes or lips, paleness, weakness, racing heart, dizziness, behavior changes and high fever.

6. **Treatment:** Treatment with influenza antivirals is recommended as early as possible for patients with confirmed or suspected influenza. Tamiflu (Oseltamivir) and Relenza (Zanamivir) are neuraminidase inhibitors. In clinical trials with seasonal influenza, these antiviral drugs have been shown to reduce the symptoms and duration of illness. Tamiflu is the best medicine to treat pregnant women who have H1N1 flu. Amentadine and Rimentadine are M2 inhibitors and can be effective for treating seasonal influenza.

7. **Supportive care:** Supportive care at home includes resting, drinking plenty of fluids and using a pain reliever for aches and pain is adequate for recovery in most cases. A non-aspirin pain reliever should be used for children or adolescents under age 18. Supportive therapy includes use of antipyretics, fluids and electrolytes.

8. **Epidemiological surveillance:** Surveillance is an important component of programmes designed to investigate the pandemics or epidemics of influenza at international level. It has three objectives such as

determination of the virus types circulating in the population at a given time so that appropriate virus variants are selected to develop vaccine; collection and analysis of morbidity/mortality data on the disease; and early detection of a new variant to evolve a system for forecasting of influenza epidemics.

Chapter 30

Zoonoses Caused by Coronaviruses

Coronaviruses belong to the family Coronaviridae. Coronaviruses are classified into three antigenic groups such as group I, II and III. Group I includes human *Coronavirus*, porcine respiratory *Coronavirus*, feline *Coronavirus* and canine *Coronavirus*, and rabbit *Coronavirus*. Group II includes another human *Coronavirus*, murine hepatitis virus, sialodacryoadenitis virus, porcine haemagglutinating encephalomyelitis virus and bovine *Coronavirus*. Group III includes infectious bronchitis virus of chickens and turkey *Coronavirus*. Until 2003, the members of the Coronaviridae have not been implicated as having zoonotic potential.

Severe Acute Respiratory Syndrome (SARS)

Severe acute respiratory syndrome is an emerging viral zoonotic disease caused by an unknown *Coronavirus* (SARS *Coronavirus*) belongs to the family Coronaviridae and characterized by respiratory problems, fever and malaise. *Coronavirus* infects many animal species and human beings also; therefore, it has zoonotic potential. SARS is the first severe infectious disease to emerge in the twenty-first century, poses a serious threat to global health security, the livelihood of populations, the functioning of health systems, and the stability and growth of economies.

Etiology

The causative agent is an unknown *Coronavirus* (SARS *Coronavirus*). SARS *Coronavirus* is intermediate between *Coronavirus* group II and III.

Epidemiology

SARS cases were first reported in Hong Kong in March 2003. Later on this disease widely distributed throughout the world. SARS *Coronavirus* had introduced to Hong Kong by a physician who had come from Guangdong, where he had treated cases of SARS in a hospital. As of late May 2003, over 8,000 cases of the SARS and over 750 deaths from this disease have been reported throughout the world among which China was most affected. As per reported cases, the case fatality rate was higher in patients older than 40 years and in those suffering from chronic diseases such as diabetes and chronic nephritis. Accordingly, case fatality rates of 3- 10 per cent or more were found.

Transmission

The principal mode of transmission is airborne infection that is transmission of infection through inhalation of infected aerosols. SARS is frequently transmitted from one person to another person. It is well established that SARS *Coronavirus* is not only transmitted via the respiratory route but also excreted in faeces. However, there is no evidence that the virus contaminates the food or toilets contribute the spread of this disease.

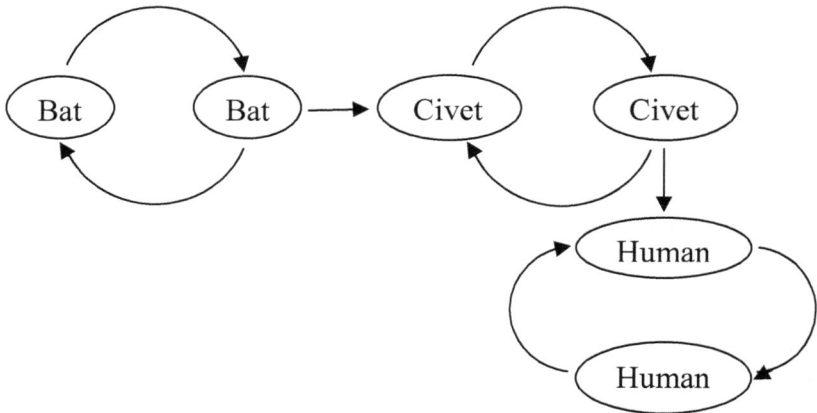

Figure 30.1: Infection Cycle of SARS.

Disease in Humans

The incubation period is about 7 days, which may vary from 4 to 12 days. Clinical manifestations of cases show a range of mild, moderate and severe respiratory disease. All clinical cases show fever and malaise. Fever may persist for up to 3 weeks, and relapse may occur. Other symptoms may be cough, dyspnea, shortness of breath, hypoxia, headache, nausea, vomiting and diarrhea. SARS cases with severe respiratory disease are characterized by radiographic evidence of atypical pneumonia and respiratory distress syndrome. On autopsy of patients with severe cases, pneumonic infiltrates and exudates are found.

Diagnosis

SARS *Coronavirus* can be isolated in mouse or human cell cultures. It can be identified in clinical specimens or cell culture material by RT-PCR. Serological tests like indirect immunofluorescence tests or Western blots are also helpful in the diagnosis. Differential diagnosis must be made with agents causing atypical pneumonia like influenza viruses, adenoviruses, *Mycoplasma*, chlamydiae, coxiellae, legionellae *etc.*

Disease Management in Humans

There is no specific antiviral drug for the treatment of SARS. Treatment requires supportive therapy especially in severe acute respiratory distress. Supplementation of oxygen is helpful to overcome the respiratory distress. No vaccine is yet available. Since the source of primary infections has not been identified, prophylactic measures are needed to prevent the spread of virus from patients to nursing staff and family members.

Chapter 31

Zoonoses Caused by Poxviruses

There is a large number of different viruses belong to the family Poxviridae. They infect wide range of vertebrate and invertebrate hosts. Poxviruses are the largest animal viruses. They are much more complex in their structures than those of any other viruses. The important animal poxviruses are *Orthopoxvirus*, *Parapoxvirus* and *Yatapoxvirus*, transmitted to humans and cause zoonotic diseases. *Parapoxvirus* is antigenically different from other poxviruses. It causes disease and produces lesions similar to those induced by other poxviruses. The important zoonotic diseases caused by these viruses are contagious ecthyma of sheep (orf), milker's nodules (pseudocowpox), cowpox, buffalopox, monkeypox *etc*.

Contagious Ecthyma of Sheep (Orf)

Synonyms

Orf, Sore mouth, Scabby mouth

Contagious ecthyma is a highly contagious skin disease commonly seen in goats and sheep. The disease is found worldwide and is caused by a *Parapoxvirus*. The disease is often more severe in goats than in sheep.

Etiology

Contagious ecthyma results from infection by the orf virus, a member of the genus *Parapoxvirus* in the family Poxviridae.

Epidemiology

Contagious ecthyma has been found worldwide in all countries that raise sheep. In the US, this disease is seen most often in the Western States. Contagious ecthyma affects sheep, goats and some other domesticated and wild ruminants. Kids and lambs are more susceptible to the disease than adults. The persons at risk of acquiring infections are shepherds, farmers, veterinarians and slaughterhouse workers. The disease outbreaks occur more frequently during periods of extreme temperatures such as late summer and winter.

Transmission

The disease is spread via direct contact with an infected animal or indirectly from a contaminated environment (feeding troughs, bedding, equipment *etc.*). In naturally occurring cases, the virus enters the hosts (sheep and goat) through abrasions on the skin of lips and face. The disease in humans is transmitted through contact with skin lesions of infected sheep. The disease may be transmitted from person to person through contact.

Disease in Animals

The disease is characterized by development of inflammation and ulcerations on the lips, mouth, vulva and around the margins of hoof. The disease may be severe which involves cauliflower-like tumors of the mouth and mucous membrane, deep ulcerative stomatitis, pharyngitis, oesophagitis and loss of hoof. The skin lesions are painful. This severe form of the disease causes high mortality in sheep.

Disease in Humans

The incubation period of the virus is 3-7 days. The disease in humans is characterized by development of lesions on the hands and arms, low grade fever and axillary lymphadenitis. Contagious ecthyma usually occurs as a single skin lesion or a few lesions. The initial lesion is a small, firm, red to blue papule at the site of virus penetration, most often a finger, hand or other exposed part of the body. The papule develops into a hemorrhagic pustule or bulla, which may contain a central crust and bleeds easily. In the later stages, the lesion develops into a nodule, which may weep fluid and is sometimes covered by a thin crust. It eventually becomes covered by a thick crust. The complications of the disease may cause toxic erythema, erythema multiforme and bullous pemphigoid.

Diagnosis

1. **Clinical examination:** The disease can be diagnosed on the basis of history of contact with sheep and clinical symptoms.
2. **Microscopic examination:** A rapid diagnosis of the disease can be made on the basis of electron microscopy of *Parapoxvirus*.
3. **Serological examination:** The suspected materials can be inoculated into cell cultures for the fluorescent antibody test (FAT).
4. **Molecular examination:** Polymerase chain reaction assays can give a definitive diagnosis.

Disease Management in Animals

1. **Hygienic measures:** Strict hygienic and sanitary measures should be applied. Cleaning and disinfecting pens may help to reduce the contamination in the environment. The best disinfectants for the poxviruses

are detergents, hypochlorite, alkalis, Virkon and glutaraldehyde. Affected animals should be isolated from rest of the flock.

2. **Quarantine measures:** New animals entering the herd should be quarantined for 3-4 weeks before mixing with other animals on the farm.

3. **Treatment:** Treatment of individually infected animals is not necessary unless lesions are severe. Severely affected kids require good nursing care to ensure that they are eating and drinking. Lesions can be treated with a single application of 3 per cent iodine solution. Animals are cured spontaneously in most cases. In severe cases of secondary bacterial infection, the usage of a systemic antibiotic is recommended. Does and ewes may require antibiotic treatment if they develop mastitis.

Disease Management in Humans

1. **Protective measures:** Avoid contact of abraded or cut skin with infected animals, scabs and crusts, wool or hides. Veterinarians and sheep handlers should exercise reasonable protective precautions and wear disposable gloves.

2. **Treatment:** In immunocompetent humans, contagious ecthyma is usually self-limiting. Treatment is supportive and typically consists of moist dressings, local antiseptics, finger immobilization and/or antibiotics to treat secondary bacterial infections. Large lesions can be removed by surgery, and curettage and electrodesiccation may be used for persistent lesions. Cryotherapy has been reported to hasten recovery.

Milker's Nodules (Pseudocowpox)

Synonyms

Pseudocowpox, Paravaccinia, Ring sore

A milker's nodule is an infection of the skin caused by a virus that infects the teats of cows.

Etiology

The disease is caused by pseudocowpox virus, a member of the genus *Parapoxvirus*.

Epidemiology

The disease has been reported from Europe and United States. The disease may occur in dairy farmers and veterinarians who examine the mouth of the infected animal.

Transmission

The virus is transmitted through direct contact from animal to animal in a herd. A breach in the skin is required for transmission of the virus. It is transmitted in milkers through contact of pox lesions on the udders of milking cows.

Disease in Animals

The disease is characterized by development of pox lesions on the udder of the cows and ulcers in the mouth of calves.

Disease in Humans

The disease in milkers is characterized by the development of papules and vesicles with inflammation of axillary lymph nodes. The pox lesions are usually seen on one hand which may extends towards arms and body.

Diagnosis

1. **Clinical examination:** The disease can be diagnosed on the basis of clinical findings.
2. **Microscopic examination:** Electron microscopy of the virus may be helpful in diagnosis of the disease.

Disease Management in Animals

1. **Isolation of animal:** Infected animals should be isolated from the herds.
2. **Disinfection:** Milking machine and teat cup should be treated with virucidal agents.
3. **Treatment:** Symptomatic treatments should be adopted. Effective teat dipping with an iodophor teat dip is probably the most effective means of control.

Disease Management in Humans

1. **Hygienic measures:** Avoid direct contact with infected cows. Milkers should wash their hands with antiseptic solutions.
2. **Protective measures:** Milker's nodules can just be left to resolve spontaneously over 4-6 weeks. They should be covered to prevent contamination of the environment and also potential spread to other people. Gloves should be worn if milking.
3. **Treatment:** Secondary bacterial infection can be treated with antibiotics.

Chapter 32

Zoonoses Caused by Alphaviruses

Alphaviruses are the members of the arboviruses (arbo: **ar**thropod-**bo**rne). Other important arboviruses are bunyaviruses and flaviviruses. The genus *Alphavirus* belong to the family Togaviridae. They are single-stranded RNA viruses and have plus-strand polarity. Alphaviruses are globally dispersed. Alphaviruses are important emerging mosquito-borne, zoonotic pathogens. They cause both localized human outbreaks and epizootics (*e.g.,* Venezuelan equine encephalitis) and large human epidemics (*e.g.,* Chikungunya). They are maintained in nature by mosquito-vertebrate-mosquito cycles. Restricted interactions between viruses, vector species, and vertebrate hosts tend to confine the geographic spread of alphaviruses. Occasionally, a virus may escape its usual ecological niche and cause widespread epizootics (*e.g.,* Venezuelan equine encephalitis virus) or urban epidemics (*e.g.,* chikungunya virus). Human infections are usually seasonal and occur in endemic areas. Alphaviruses cause important zoonoses like Eastern equine encephalitis, Western equine encephalitis, Venezuelan equine encephalitis, chikungunya fever, Semliki forest fever, Ross river fever, Sindbis fever, O'nyong fever, Mayaro fever *etc.* The clinical manifestations due infections caused by alphaviruses range from frank, severe encephalitis (*e.g.,* Eastern and Western equine encephalitis) to polyarthritis (*e.g.,* Ross River fever).

Chikungunya Fever

Chikungunya fever is a mosquito-transmitted acute febrile infection caused by *Alphavirus.* The name chikungunya comes from Swahili and means, "what bends". In humans it causes severe arthralgia and a maculopapular rash. Chikungunya fever is clinically inapparent in birds.

Etiology

The causative agent of chikungunya fever is *Alphavirus* belong to the family

Togaviridae. The reservoirs of the causative agent are primates, bats, birds *etc.* The antibodies against infection have been found in green monkeys, orangutans and chimpanzees. This virus is serologically closely related to O'nyong-nyong, Myaro, Ross River and Semliki Forest viruses.

Epidemiology

Chikungunya fever was first reported in Tanzania in 1953. Later on the disease has been reported from other parts of Africa and Asia. Antibodies against this virus have been reported in monkeys, orangutans, chimpanzees, rodents and birds from many countries such as Tanzania, Uganda, Zimbabwe, Zaire, South Africa and Southeast Asia. The disease is epidemic in nature. Chikungunya virus has been reported from Bangkok in 1960s; various parts of India including Vellore, Calcutta and Maharashtra in 1964; in Sri Lanka in 1969; Vietnam in 1975; Myanmar in 1975 and Indonesia in 1982.

In India, the epidemics of this disease have been reported in humans in Kolkata and Chennai. Kolkata strain is found to be more virulent than Chennai, Nagpur or African strain of the virus. The virus has been recovered from vector of this infection. In 2005, an outbreak of chikungunya involving 258,000 cases with 219 deaths was recorded on the French island of Reunion in the Indian Ocean. In 2006, several outbreaks of chikungunya fever have been reported from various states/ provinces of India such as Maharashtra, Gujarat, Madhya Pradesh, Delhi, Andhra Pradesh, Tamil Nadu, Karnataka, Kerala, Andaman and Nikobar. In India, recent outbreaks of chikungunya fever have also been reported in 2015 and 2016.

Transmission

The disease is transmitted by mosquito vectors. In Asia and Africa, the important vectors are *Aedes aegypti* and *Aedes africanus*, respectively. The infection is maintained through urban, rural and rain-forest cycles. Human beings are the main source of infection for mosquitoes. Therefore, urban outbreaks of disease involve human-to-human transmission by mosquitoes. There is no vertical transmission of virus in mosquitoes. The disease is transmitted through infected mosquito bite. *Aedes aegypti* has been incriminated as the main vector. Other vectors such as *Aedes africanus, Aedes furcifer taylori, Culex pipiens fatigans* and *Mansonia* species have also been involved in the epidemics of this disease in African countries. The virus is maintained in two cycles that is urban cycle and sylvan cycle. Urban cycle is maintained between man and mosquito, while sylvan cycle in wild animals and mosquito. Chikungunya shows a cyclical pattern and epidemic occurs every 4-6 years.

Clinical Manifestations

The incubation period of chikungunya virus is about 6-10 days. The disease is characterized by sudden onset of fever and arthralgia, which immediately causes patients to unable to move. The fever is usually biphasic. The maculopapular rashes may develop around 2-5 days, which can be hemorrhagic. Arthralgia is often clinically manifested by swelling, redness and sensitivity to pressure of joints. Arthralgia and edema may persist for weeks after the acute disease. Back

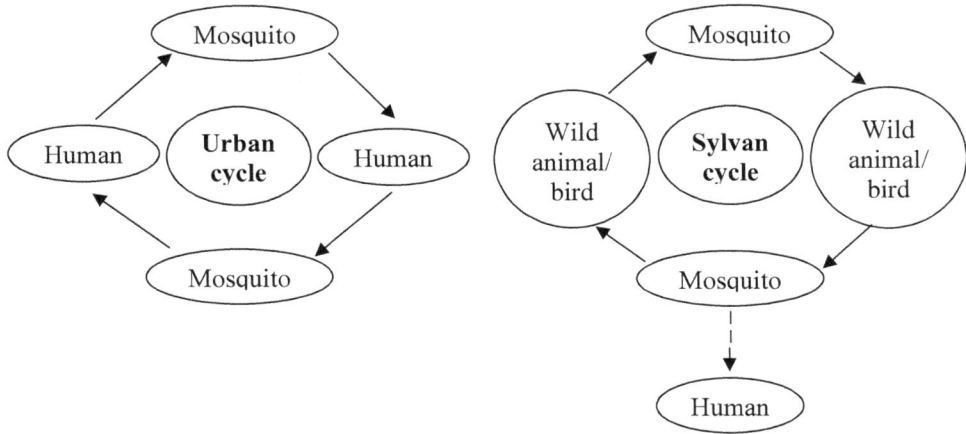

Figure 32.1: Infection Cycles of Chikungunya Fever.

and limb pain may be severe. Other symptoms may be myalgia, retroocular pain, conjunctivitis, photophobia, lymphadenopathy, headache, nausea and vomiting. Hemorrhagic complications have not been reported in African epidemics, while have been observed in Asia in 5-7 per cent cases.

Diagnosis

1. Isolation of the virus from blood of patient after 6 days of first symptoms.
2. Detection of the virus, using RT-PCR technique.
3. Differential diagnosis with dengue fever (No arthritis but hemorrhagic manifestations are common). Arthritis is common, while hemorrhagic manifestations are uncommon in chikungunya fever.

Disease Management in Humans

1. **Control of mosquitoes:** There are several methods of mosquito control but experts recommends an "integrated approach", that is an approach which avoids the excessive use of any one method but tries to combine two or more methods with a view to obtain maximum results with minimum inputs and also prevent environmental pollution with toxic chemicals and development of insecticide resistance. The various methods of mosquito control are as follows:

   Environmental control: It includes elimination of breeding places, planned water management *etc.* The environment should be cleaned up and got rid of water holding containers such as discarded tins, empty pots, broken bottles, and coconut shells and similar other artificial collection of water.

   Chemical control: Adult mosquitoes are most commonly controlled by spraying the houses with residual insecticides. In space spray the insecticides are sprayed into the atmosphere in the form of a mist or fog to kill the mosquitoes. The most common space spray is pyrethrum extract, which can be sprayed at a dosage of 1 oz of spray solution per 1,000 cubic feet of space. Mineral oils such as diesel oil, fuel oil, kerosene and various fractions of crude oils can be used on water to kill the larvae and pupae within a short time of application. The usual rate of oil application is 40 to 90 litres per hectare. Oil should be applied once a week (as life cycle of mosquito occupies about 8 days). Paris green is can be applied at rate of 1 kg of actual paris green per hectare of water surface. Synthetic insecticides such as fenthion, chlorpyrifos and abate are the most effective larvicides. Abate at a concentration of 1.0 ppm has been found to be very effective larvicide.

Herbal control: The extracts of many plants of Indian origin being used for control of insect pest of plants, animals and humans. Neem (*Azadirachta indica*) has been successfully used for control of mosquitoes.

Biological control: Some larvivorous fish such as *Gambusia, Lebister reticulatus* (sometimes known as Barbados Millions) and *Poecilia reticulata* (guppy fishes) can be used effectively against mosquito larvae. These fish can be used in burrow pits, sewage oxidation ponds, ornamental ponds, cisterns and farm ponds. There are several reports of use of some microorganisms against mosquito control. *Bacillus thuringiensis* var *israelensis, Bacillus sphaericus* and pathogenic microsporidian such as *Thelokomia opecita* have been used for control of mosquitoes.

Genetic control: In recent years, mosquito control by genetic methods such as sterile male technique, cytoplasmic incompatibility, chromosomal translocation, sex distortion, and gene replacement have been explored. They have certain advantages over chemical methods, being cheaper and potentially more efficient and not subjected to vector resistance.

2. **Protection against mosquito bites:** It includes the use of mosquito net and mosquito repellents. The mosquito net offers protection against mosquito bites during sleep. There should not be a single hole or rent in the net. The size of the openings in the net should not exceed 0.0475 inch in any diameter. There should be provision of screening in the buildings. The aperture should not be larger than 0.0475 inch. There are many repellents such as diethyl toluamide, ethyl hexanediol, dimethyl phthalate, dimethyl carbate, and indalone, which can be used against mosquitoes. Diethyl toluamide (deet) has been found to be an outstanding all- purpose repellent.

3. **Vaccination:** For human beings the commercial vaccine is not available. An experimental attenuated live virus vaccine is available for exposed laboratory personnel.

4. **Treatment:** Treatment of chikungunya fever is symptomatic. Strong analgesic is required in severe joint pain. Use of steroid must be avoided.

Chapter 33

Zoonoses Caused by Flaviviruses

Flaviviruses are important members of the Arboviruses. The genus *Flavivirus* belongs to the family Flaviviridae. Flaviviruses are enveloped RNA viruses with a cubic nucleocapsid and a single-stranded genome with plus polarity. They are maintained in nature by transmission in mosquito-vertebrate-mosquito or tick-vertebrate-tick cycles. With yellow fever and dengue viruses, humans are important intermediate hosts during urban epidemics. Human infections are seasonal and are acquired in endemic areas. Infection is initiated by the bite of an infected mosquito or tick. Virus disseminates during lytic infection of cells, causing viremia. Flaviviruses cause syndromes like encephalitis (St. Louis encephalitis, Japanese encephalitis, Powassan, and tick-borne encephalitis), febrile illness with rash (dengue fever), hemorrhagic fever (Kyasanur forest disease and sometimes dengue fever), and hemorrhagic fever with hepatitis (yellow fever).

Dengue Fever

Dengue fever (DF) is the most common mosquito-borne viral disease of humans that in recent years has become a major international public health concern. The global incidence of dengue has grown dramatically in recent decades. About half of the world's population is now at risk. Dengue is found in tropical and sub-tropical climates worldwide, mostly in urban and semi-urban areas. Globally, 2.5 billion people live in areas where dengue viruses can be transmitted. Severe dengue is a leading cause of serious illness and death among children in some Asian and Latin American countries.

Etiology

Etiology of the disease is associated with several aspects that are etiological agent, reservoir and vector. Dengue is caused by *Flavivirus* (an arbovirus). There are four serotypes of dengue virus, namely, DEN-1, DEN-2, DEN-3 and DEN-4.

Recovery from infection by one provides lifelong immunity against that particular serotype. However, cross-immunity to the other serotypes after recovery is only partial and temporary. Subsequent infections by other serotypes increase the risk of developing severe dengue. The reservoirs of infection are man and mosquito. Monkeys are susceptible to this virus. Antibodies against DEN-2 virus have been found in buffalo, goat, horse swine and dogs. The main vector of the dengue syndrome is *Aedes aegypti*. Outbreaks of dengue have also been attributed to *A. albopictus, A. polynesiensis* and several species of the *A. scutellaris complex*.

Epidemiology

Severe dengue also known as dengue hemorrhagic fever (DHF) was first recognized in the 1950s during the dengue epidemics in the Philippines and Thailand. By 1970 nine countries had experienced epidemic of DHF and now, the number has increased more than fourfold and continues to rise. Today emerging DHF cases are causing increased dengue epidemics in America and Asia, where all four dengue viruses are endemic. A pandemic in 1998, in which 1.2 million cases of dengue fever (DF) and dengue hemorrhagic fever (DHF) were reported from 56 countries was unprecedented. Currently DF/DHF is endemic in India, Indonesia, Bangladesh, Maldives, Myanmar, Thailand and Sri Lanka and approximately 1.3 billion people are living in the endemic areas. Dengue is Notifiable disease in Sri Lanka, Indonesia, Thailand and Myanmar. The increase in dengue and DHF is due to uncontrolled population growth and urbanization without appropriate water management, the global spread of dengue via travel and trade, and to erosion of vector control programme.

Transmission

Dengue virus is transmitted by infected mosquito bite. Once the mosquito becomes infective, it remains so far life. Transovarian transmission has been demonstrated in laboratory. The virus is maintained in two cycles that is urban cycle and sylvan cycle. Urban cycle is maintained between human and mosquito, while sylvan cycle in monkey and mosquito. Dengue virus is transmitted by female mosquitoes mainly of the species *Aedes aegypti* and, to a lesser extent, *Aedes albopictus*. The *Aedes aegypti* mosquito lives in urban habitats and breeds mostly in man-made containers. Unlike other mosquitoes *Aedes aegypti* is a day-time feeder; its peak biting periods are early in the morning and in the evening before dusk. Female *Aedes aegypti* bites multiple people during each feeding period. *Aedes albopictus*, a secondary dengue vector in Asia, has spread to North America and Europe largely due to the international trade in used tyres (a breeding habitat) and other goods (*e.g.*, lucky bamboo). *Aedes albopictus* is highly adaptive and, therefore, can survive in cooler temperate regions of Europe. Its spread is due to its tolerance to temperatures below freezing, hibernation, and ability to shelter in microhabitats. After virus incubation for 4–10 days, an infected mosquito is capable of transmitting the virus for the rest of its life. Infected humans are the main reservoirs and multipliers of the virus, serving as a source of the virus for uninfected mosquitoes.

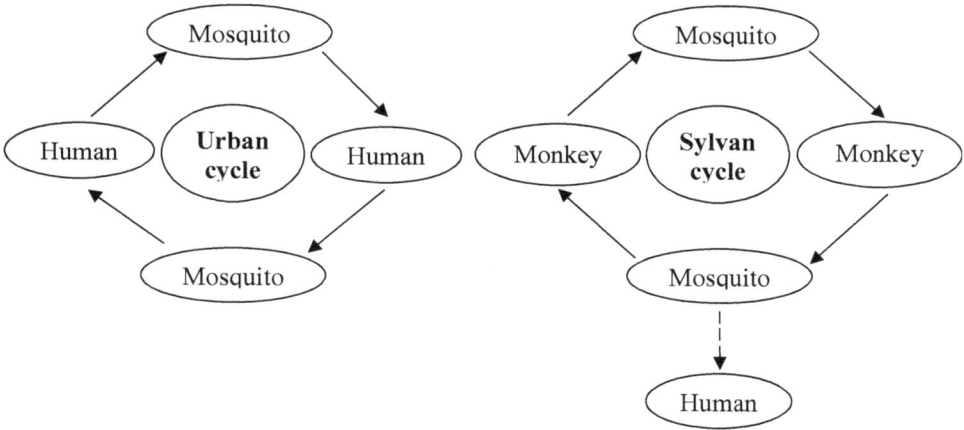

Figure 33.1: Infection Cycles of Dengue Fever.

Disease in Humans

The disease in humans may be asymptomatic or may lead to classical DF, DHF or dengue shock syndrome (DSS).

1. **Classical dengue fever (DF):** It is also known as "break-bone fever" in India for a very long time. It is an acute viral infection caused by at least 4 serotypes of dengue virus. All ages and both sexes are susceptible to dengue fever. Children have usually milder disease than adults. The disease is characterized by sudden onset of high fever, chills, intense headache, myalgia, arthralgia, retroocular pain, colic, dragging pain in inguinal region, sore throat, general depression *etc.* The skin eruption appears in 80 per cent of cases during the remission or during second febrile phage, which last for 1-2 days. The rashes accompanied by similar but milder symptoms. Fever lasts for about 5 days, rarely more than 7 days after which recovery is usually complete although convalescence may be protracted. The case fatality is exceedingly low.

2. **Dengue hemorrhagic fever (DHF):** Severe dengue (also known as dengue hemorrhagic fever) is characterized by fever, abdominal pain, persistent vomiting, bleeding and breathing difficulty and is a potentially lethal complication, affecting mainly children. It is caused by infection with more than one serotype of dengue virus. Hepatomegaly, petechiae, shock and bleeding are also common.

3. **Dengue shock syndrome (DSS):** This is characterized by all criteria of DHF plus shock manifested by rapid and weak pulse, hypotension with the presence of cold clammy skin and restlessness.

Diagnosis

1. **Clinical examination:** The common clinical manifestations are acute onset of high fever and lasting 2-7 days. Hemorrhagic manifestations such as

petechiae, perpura, echymosis, epistaxis, gum bleeding, hemoptysis and or melena are also common.

2. **Laboratory examination:** The clinical criteria such as thrombocytopenia (1 lac/mm^3 or less), hemoconcentration and increased hematocrit by 20 per cent or more are sufficient to establish a clinical diagnosis of DHF.

Disease Management in Humans

1. **Control of mosquitoes:** At present, the only method to control or prevent the transmission of dengue virus is to combat vector mosquitoes through preventing mosquitoes from accessing egg-laying habitats by environmental management and modification; introduction of natural enemies like larvivorous fishes into mosquito habitat; disposal of solid wastes properly and removing man-made habitats; covering, emptying and cleaning of domestic water storage containers on a weekly basis; use of personal household protection such as window screens, long-sleeved clothes, use of mosquito nets, coils and vaporizers; mosquito repellents; improving community participation and mobilization for sustained vector control; and applying insecticides as space spraying during outbreaks.

2. **Vaccination:** No immediate prospect of preventing the disease by immunization. There is no vaccine to protect against dengue. However, major progress has been made in developing a vaccine against dengue/severe dengue. Three tetravalent live-attenuated vaccines are under development in phase II and phase III clinical trials, and 3 other vaccine candidates (based on subunit, DNA and purified inactivated virus platforms) are at earlier stages of clinical development. The WHO provides technical advice and guidance to countries and private partners to support vaccine research and evaluation.

3. **Surveillance:** Strengthening epidemiological surveillance through the implementation of "DengueNet" developed by the WHO is an important strategy for prevention and control of dengue. Moreover, active monitoring and surveillance of vectors should be carried out to determine effectiveness of control interventions.

4. **Treatment:** There is no specific treatment for dengue/severe dengue, but early detection and access to proper medical care lowers fatality rates below 1 per cent. Management of dengue fever is symptomatic and supportive. The management of DHF during the febrile phage is similar to that of DF. A rise in hematocrit value indicates significant plasma loss and a need for parentral fluid therapy. DSS is a medical emergency that requires prompt and vigorous volume replacement therapy.

5. **Miscellaneous measures:** Isolation of patient under bed nets during first few days and personal protection like covering of exposed skin with clothing at dusk, use of mosquito nets *etc.* may be some measures.

Japanese Encephalitis (JE)

Japanese encephalitis (JE), formerly called Japanese B encephalitis, is caused by a *Flavivirus* and transmitted by mosquitoes. It is a viral zoonotic disease, infecting mainly animals and accidentally humans. Twenty four countries in the WHO Southeast Asia and Western Pacific regions have endemic Japanese encephalitis virus (JEV) transmission, exposing more than 3 billion people to risks of infection.

Etiology

The disease is caused by *Flavivirus* (group B arbovirus), which is closely related to the agents of St. Louis encephalitis virus, West Nile fever and Murray Valley encephalitis viruses and transmitted by culicine mosquitoes such as *Culex tritaeniorhynchus, C. vishnui* and *C. gelidus* along with some anophelines. These mosquitoes generally breed in irrigated rice fields, shallow ditches and pools. Among these, *C. tritaeniorhynchus* has been implicated as the most important vector in India.

Epidemiology

Japanese encephalitis has emerged as a major public health problem. It was first reported in 1935 in Japan. About thirty years ago JE was known as an endemic disease in East Asia especially Japan, China and Korea, but in recent years the disease has spread widely in Southeast Asia. JEV is the main cause of viral encephalitis in many countries of Asia with an estimated 68 000 clinical cases every year. The outbreaks of considerable magnitude have occurred in India, Sri Lanka, Thailand, Indonesia, Viet Nam and Myanmar. The outbreaks between 1986 and 1990, the registered cases of JE in China, India, Thailand and Japan were 126000, 26000, 836 and 122, respectively. An outbreak of JE occurred on the island of Spain for the first time in 1990. Between 1989 and 1993 three tourists became infected in Bali and developed JE, and one tourist died.

The spread of JE in Australia is limited due to the fact that 80 per cent of wild boars have antibodies against Murray Valley encephalitis virus and Kunjin virus. These antibodies protects against infection with JE virus. JE occurs frequently in Malaysia but not in Indonesia. The reduced pig farming in the Islamic country may be one of the reasons. The disease is rare in other parts of the world and when seen, is generally associated with travelers returning from endemic areas. Up to 70 per cent of adults in tropical region of Asia have antibodies against JE virus. Transovarian transmission of JE virus in mosquitoes has been proven. They, therefore, can serve as virus reservoirs for over winter survival even in areas with a temperate climate. In endemic areas the virus can be isolated from the mosquitoes. The occurrence of JE is higher in rainy season due to propagation of mosquitoes.

In India, Japanese encephalitis was reported in 1955 in Tamil Nadu. Subsequent surveys carried out by the National Institute of Virology, Pune indicated that about half of the population of South India had neutralizing antibodies to this virus. JE has been reported time to time from India but in the last few decades, there has been a major upsurge of JE in most of the states of India such as U.P., M.P., A.P., Karnataka, Tamil Nadu, Bihar, West Bengal, Assam, Goa, Pondicherry and Maharashtra. For the last two decades the outbreaks of Japanese encephalitis in India have been reported almost every year. JE is endemic in U.P. and some south Indian stases like Andhra Pradesh and Tamil Nadu.

The risk of JE is very low for most travellers to Asia, particularly for short-term visitors to urban areas. However, the risk varies according to season, destination, duration of travel and activities. Vaccination is recommended for travellers with extensive outdoor exposure (camping, hiking, working, *etc.*) during the transmission season, particularly in endemic countries or areas where flooding irrigation is practiced. JE primarily affects children. Most adults in endemic countries have natural immunity after childhood infection, but individuals of any age may be affected. Prevalence is generally higher in males and fatalities varies from region to region such as 10 per cent China, 30 per cent Japan, 21-40 per cent India and more than 40 per cent Korea. The spread of the disease correlates well with the densities of mosquito vectors. It has been seen that outbreaks generally occur during summer and autumn months.

Transmission

Pigs and birds such as egrets and pond herons have been incriminated as the most important hosts for maintenance of JE virus. JE is transmitted by mosquito vectors. Human infection occurs through infected mosquito bite. The main vector of JE virus is *C. tritaeniorhynchus*, which hatches in the rice fields. Its multiplication is increased by 50 per cent in fertilized fields. The increase in JE cases in India may be due to increased use of fertilizers in the rice fields. Transmission occurs principally in rural agricultural locations where flooding irrigation is practiced, some of which may be near or within urban centres. Transmission is related mainly to the rainy season in Southeast Asia but may take place all year round, particularly in tropical climate zones. In the temperate regions of China, Japan, the Korean peninsula and Eastern parts of the Russian Federation, transmission occurs mainly during the summer and autumn. The seasonal occurrence of JE in northern region is caused by migrating birds especially herons. Human is an "accidental dead end host". Transmission from human to human or animal to animal has not been reported. Humans, once infected, do not develop sufficient viraemia to infect feeding mosquitoes. The virus exists in a transmission cycle between mosquitoes, pigs and/or water birds (enzootic cycle).

Disease in Animals

Under natural conditions, several vertebrates such as pigs, cattle, goats, cats, dogs, birds, bats, snakes, and toads are infected. Among the animals, pigs have been incriminated as the major vertebrate host for JE virus. In some places, up to 100 per cent of pigs may be infected with JE virus. Infected pigs generally do not manifest any overt symptoms of illness but circulate the virus so that mosquitoes get infected and can transmit the virus to humans. Occasionally JE virus causes abortion in pigs. The pigs are considered as "amplifier host" of the virus. Cattle and buffaloes may also be infected with JE virus, although they may not be natural host of JE virus. They act as "mosquito attractants". Horses infected with JE virus manifest signs of encephalitis. Some species of birds such as pond herons, poultry and ducks appear to be involved in the natural history of JE virus.

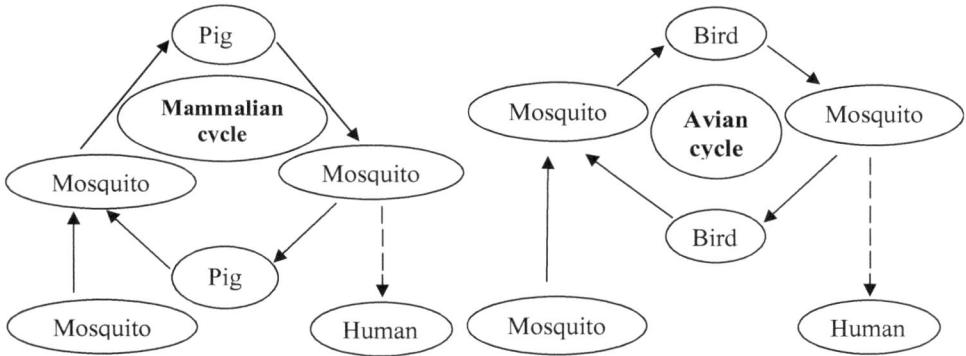

Figure 33.2: Infection Cycles of Japanese Encephalitis.

Disease in Humans

In humans, most JE virus infections are mild (fever and headache) or without apparent symptoms, but approximately 1 in 250 infections results in severe clinical illness. Severe disease is characterized by rapid onset of high fever, headache, neck stiffness, disorientation, coma, seizures, spastic paralysis and ultimately death. The case-fatality rate can be high (up to 40 per cent) among those with disease symptoms. Of those who survive, 20 per cent –30 per cent suffer permanent intellectual, behavioural or neurological problems such as paralysis, recurrent seizures or the inability to speak. The incubation period is 4-14 days. The disease in humans occurs in three stages-

1. **Prodromal stage:** The onset of illness is usually acute and heralded by fever, headache, malaise and non-specific respiratory and gastrointestinal disturbances. The duration of this stage is 1-6 days.

2. **Acute encephalitis stage:** Fever is usually high (100-105°F). The prominent features are fever, nuchal rigidity, focal CNS signs, convulsion, and altered sensorium, which may lead to coma.

3. **Late stage and sequelae:** Neurological signs become stationary or tend to improve. The case fatality rate varies between 20- 40 per cent. The average period between the onset of illness and death is about 9 days.

Diagnosis

Individuals who live in or have travelled to a JE-endemic area and experience encephalitis are considered a suspected JE case. To confirm JEV infection and to rule out other causes of encephalitis requires a laboratory testing of serum or, preferentially, cerebrospinal fluid. Surveillance of the disease is mostly syndromic for acute encephalitis. Confirmatory laboratory testing is often conducted in dedicated sentinel sites, and efforts are undertaken to expand laboratory-based surveillance. Case-based surveillance is established in countries that effectively control JE through vaccination.

1. **Tentative diagnosis:** It can only be made in connection with the epidemiological situations (history of traveling or epidemiology).

2. **Serological diagnosis:** It can be made by demonstration of virus specific IgM antibodies in the CSF.

3. **Molecular diagnosis:** More recently, RT- PCR has been recommended to confirm the diagnosis in postmortem, but it has not been recommended in ante-mortem diagnosis. PCR can also be used to detect the virus in mosquitoes for disease control.

4. **Isolation of virus:** Diagnosis can also be made by isolation of JE virus from brain tissues of affected individuals. In epidemic areas the virus can also be isolated from the mosquitoes.

Disease Management in Animals

Disease in the animals can be managed by control of mosquitoes, vaccination of the animals (pigs) and symptomatic treatment of the infected animals. Vaccine is available for both swine and horses. Two types of vaccines (modified live and inactivated) are prepared. Vaccination in swine prevents reproductive disorders and directly impacts JE viral amplification, especially in enzootic areas.

Disease Management in Humans

Japanese encephalitis can be controlled by interruption of transmission cycle between reservoir/amplifier host, biological vectors and susceptible population. Japanese encephalitis can be prevented and controlled by the following measures-

1. **Control of mosquitoes:** Mosquitoes can be controlled by using insecticides, larvicidal fish in water logging areas and proper water management. Growing of the water fern *Azolla microphylla* in rice field has been found as a biological agent against mosquitoes breeding in rice fields. But long term, cost-effective and eco-friendly technology to reduce/eliminate JE vectors are not within scope so far. Fogging with ultra low volume (ULV) insecticides such as malathion and fenitrithion is helpful in reducing the mosquito vector population. Moreover, the public should be encouraged to avoid the exposure to mosquitoes by using mosquito net and mosquito repellents.

2. **Change in farming practices:** Japanese encephalitis can be successfully controlled by keeping rice crops and pig farming in separate locations. In Japan and Korea, the JE has been reduced due changes of pig farming.

3. **Vaccination:** Safe and effective JE vaccines are available to prevent disease. The WHO recommends the strong JE prevention and control activities, including JE immunization in all regions where the disease is a recognized public health priority, along with strengthening surveillance and reporting mechanisms. Even if the number of JE-confirmed cases is low, vaccination should be considered where there is a suitable environment for JE virus transmission. The WHO recommends that JE vaccination be integrated into national immunization schedules in all areas where JE

disease is recognized as a public health issue. There are 4 main types of JE vaccines currently in use *viz.*, inactivated mouse brain-derived vaccines, inactivated Vero cell-derived vaccines, live attenuated vaccines, and live recombinant vaccines. Over the past years, the live attenuated SA14-14-2 vaccine manufactured in China has become the most widely used vaccine in endemic countries, and it was prequalified by WHO in October 2013. Cell-culture based inactivated vaccines and the live recombinant vaccine based on the yellow fever vaccine strain have also been licensed and WHO-prequalified. Immunization with killed mouse brain vaccine (strain –Beijing 1 of JE virus) can be carried out as follows:

Primary immunization: 2 doses of 1ml of each should be administered subcutaneously at an interval of 7-14 days. The dose is 0.5 ml for children under the age of 3 years.

Booster dose: 1ml of vaccine should be administered subcutaneously after few months (before 1 year). Protective immunity develops in about a month's time after the second dose.

Revaccination: Revaccination is to be carried out every 3 years.

Note: JE vaccine containing Nakayama strain of the virus has been found very effective in preventing JE incidence in Japan.

4. **Treatment:** There is no antiviral treatment for patients with JE. Treatment is supportive to relieve symptoms and stabilize the patient. Anticonvulsive drugs are used for seizures and mannose infusions are given in case of increased intracranial pressure.

5. **Other measures:** All travellers to JE endemic areas should take precautions to avoid mosquito bites to reduce the risk for JE. Travellers spending extensive time in JE endemic areas are recommended to get vaccinated. Personal preventive measures include the use of repellents, long-sleeved clothes, coils and vaporizers are also important preventive measures.

West Nile Fever (WNF)

West Nile fever is a vector-borne viral zoonotic disease. The disease can cause fatal neurological symptoms in humans. However, approximately 80 per cent infected people do not show any symptoms. West Nile virus (WNV) can cause severe disease and death in horses.

Etiology

West Nile fever is caused by West Nile virus (WNV), a member of the genus *Flavivirus*. It belongs to the Japanese encephalitis antigenic complex of the family Flaviviridae.

Epidemiology

West Nile fever is commonly found in Africa, Europe, the Middle East, North America and West Asia. In India, the disease has been reported from Karnataka and Uttar Pradesh. Birds are the natural hosts of WNV. The virus is maintained in nature in a cycle involving transmission between birds and mosquitoes (*Culex* spp.). Humans, horses and other mammals can be infected with WNV. The virus was first isolated in a woman in the West Nile district of Uganda in 1937. It was identified in birds (crows and columbiformes) in Nile delta region in 1953. Before 1997 WNV was not considered pathogenic for birds, but at that time in Israel a more virulent strain caused the death of different bird species presenting signs of encephalitis and paralysis. Human infections attributable to WNV have been reported in many countries in the world for over 50 years. In 1999, a WNV circulating in Israel and Tunisia was imported in New York producing a large and dramatic outbreak that spread throughout the continental United States of America in the following years. The WNV outbreak in USA (1999-2010) highlighted that importation and establishment of vector-borne pathogens outside their current habitat represent a serious danger to the world. The largest outbreaks occurred in Greece, Israel, Romania, Russia and USA. Since its introduction in 1999 into USA, the virus has spread and is now widely established from Canada to Venezuela.

Transmission

1. **Mosquito bites:** West Nile virus is mainly transmitted to people through infected mosquito bites. The important mosquito vectors are *Culex* spp. The principal vector of WNV is *Culex univittatus* in epidemics in southern Africa. Mosquitoes become infected when they feed on infected birds, which circulate the virus in their blood for a few days.

2. **Contact:** The virus may also be transmitted through contact with other infected animals, their blood, or other tissues.

3. **Other routes:** A very small proportion of human infections have occurred through organ transplant, blood transfusions and breast milk. Transmission of WNV to laboratory workers has been reported. There is one reported case of transplacental WNV transmission. However, human-to-human transmission of WNV through casual contact has not been documented till date, and no transmission of WNV to health care workers has been reported when standard infection control precautions have been put in place.

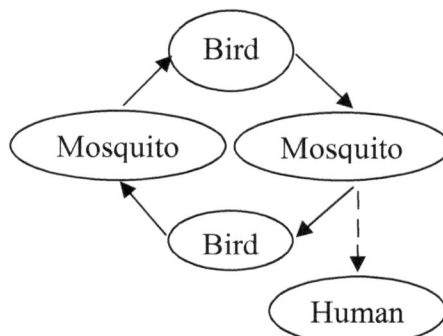

Figure 33.3: Infection Cycle of WNF.

Disease in Animals

The WNV can cause infection in birds, swine, sheep, cattle and dogs. However, the infection remains latent or exhibits only a mild severity. The virus can cause encephalitis in horses, cattle and other mammals. The virus causes fatal disease in birds, especially in crows and even in alligators. In birds, the prominent symptoms are encephalitis and paralysis.

Disease in Humans

The incubation period is usually 3 to 14 days. The infection is asymptomatic in approximately 80 per cent people. However, the infection in approximately 20 per cent people can cause severe disease. The clinical manifestation of the disease includes fever, headache, tiredness, and body aches, nausea and vomiting. Occasionally, there may be a skin rash especially on the trunk of the body and swollen lymph glands. The clinical manifestations of severe disease (also called neuroinvasive disease, such as West Nile encephalitis or meningitis or West Nile poliomyelitis) include high fever, headache, muscle weakness, nuchal rigidity, stupor, disorientation, coma, tremors, convulsions and paralysis.

Diagnosis

1. **Virus isolation:** The virus can be isolated from the blood of patients on the first day of the disease and all through the febrile phase with viremia.

2. **Serological examination:** IgG antibody seroconversion (or significant increase in antibody titers) in two serial specimen collected at a one week interval by ELISA, IgM antibody capture ELISA and neutralization assays. IgM can be detected in nearly all cerebrospinal fluid and serum specimens received from WNV infected patients at the time of their clinical presentation. Serum IgM antibody may persist for more than a year.

3. **Molecular examination:** The virus can be detected by using reverse transcription polymerase chain reaction (RT-PCR) assay.

Disease Management in Animals

The disease can be prevented and controlled by careful handling of infected birds and control of mosquitoes. Care should be taken as the birds can be infected by a variety of routes other than mosquito bites, and different species may have different potential for maintaining the transmission cycle. Horses, just like humans, are "dead-end" hosts, that is infected horses do not spread the infection. Vaccines are available for use in horses. Treatment is supportive and consistent with standard veterinary practices for animals infected with a viral agent.

Disease Management in Humans

1. **Surveillance:** An active animal health surveillance system to detect new cases in birds and horses is essential in providing early warning for veterinary and human public health authorities. In the Americas, it is important to help the community by reporting dead birds to local authorities.

2. **Vector Control:** Effective prevention of human WNV infections depends on the development of comprehensive, integrated mosquito surveillance and control programmes in areas where the virus occurs. Emphasis should be on integrated control measures including source reduction (with community participation), water management, chemicals, and biological control of vectors.

3. **Public health education:** Vaccines are not available for humans; therefore, raising awareness among people about risk factors can be effective measure for prevention of the disease. Public health educational messages should focus on reducing the risk of mosquito transmission. Efforts to prevent transmission should first focus on personal and community protection against mosquito bites through the use of mosquito nets, insect repellent, protective clothing (light coloured, long-sleeved shirts and trousers) and by avoiding outdoor activity at peak biting times. Moreover, community programmes should encourage people to destroy mosquito breeding sites in residential areas.

4. **Protective measures:** In order to reduce the risk of animal-to-human transmission, gloves and other protective clothing should be worn while handling sick animals or their tissues, and during slaughtering and culling procedures. To reduce the risk of transmission through blood transfusion and organ transplant, the blood and organ donation restrictions and laboratory testing should be considered at the time of the outbreak in the affected areas. Samples taken from humans and animals with suspected WNV infection should be handled by trained staff working in fully equipped laboratories.

5. **Treatment:** Treatment is supportive for patients with neuro-invasive West Nile virus, include hospitalization, intravenous fluids, respiratory support, and prevention of secondary infections.

Kyasanur Forest Disease (KFD)

Synonyms

Haemorrhagic fever, Monkey disease

Kyasanur forest disease is a febrile illness associated with hemorrhages caused by *Flavivirus* and transmitted to humans by infected tick bites. The disease was first recognized in 1957 in Shimoga district of Karnataka state, India. Local inhabitants called the disease "monkey disease" because of its association with death of monkeys. The disease was later named after the locality Kyasanur forest from where the virus was first isolated.

Etiology

Kyasanur forest disease is caused by *Flavivirus*, a member of group B Togavirus. It is antigenically related to Alkhurma virus of Saudi Arabia and other tick-borne *Flavivirus*. Small mammals particularly rodents and squirrels are the main reservoir of the virus. Monkeys are recognized as "amplifier host" for the virus. However they are not effective maintenance host because most of them die from KFD infection. Cattle are very important in maintaining tick population but play no part in virus maintenance (cattle provide *Haemaphysalis* ticks with a plentiful source of blood meal, which leads to population explosion among the ticks). Man is an incidental or dead end host and plays no part in virus transmission. Ticks such as *Haemaphysalis spinigera* and *H. turtura* are important vector of KFD virus.

Epidemiology

Kyasanur forest disease occurs in Shimoga, Chikamangaloor (formerly Mysore) and some northern and southern districts of Karnataka. Serological survey in different parts of India revealed antibodies to KFD or closely related virus in animals and human beings particularly in Kutch and Saurashtra. Death of monkeys is considered as heralder of this disease in endemic areas. The outbreak during 1983-84 seems to be the largest outbreak involved 69 deaths among 2167 cases. In human beings majority of cases affected were between 20-40 years of age. Attack rate was higher in males than in females. The attacked people were mostly cultivators who visited forests accompanying their cattle or cutting wood. The epidemic period correlates well with the period of greatest human activity in the forest that is January until the onset of rains in June. The highest number of human and monkey infection occurs during drier months particularly from January to June, this period coincides with the peak nymphal activity of ticks. The number of reported human cases is between 100- 500 per year.

Transmission

The important reservoirs of KFD virus are monkey, small rodents and birds. The transmission cycle of KFD involves mainly monkeys and ticks. The disease is transmitted by the bite of infected ticks, particularly nymphal stages. There is no evidence of transmission of infection from human-to-human.

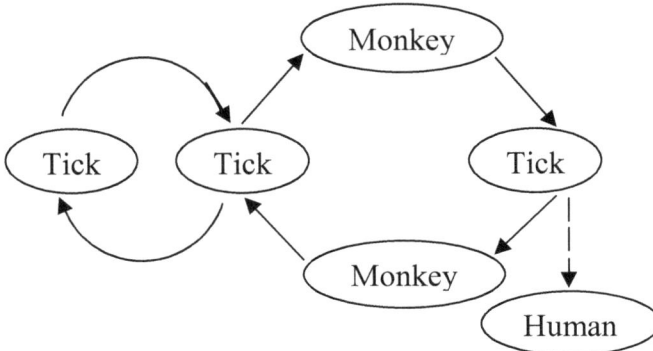

Figure 33.4: Infection Cycle of KFD.

Disease in Animals

KFD virus causes encephalitis and hemorrhage in humans and animals. But clinical manifestations are best seen in human cases. Monkey may show encephalitis. High mortality rate occurs in monkeys.

Disease in Humans

Clinical manifestations are sudden onset of fever, headache, severe myalgia and prostration. In severe cases, there may be gastrointestinal disturbances; hemorrhages from nose, gums, stomach and intestine. In most of the cases, a second phase of infection also occurs which is characterized by mild meningoencephalitis and manifested by a return of fever, severe headache followed by stiffness of the neck, tremors, abdominal reflexes and mental disturbances. The case fatality rate has been estimated to be 5-10 per cent.

Diagnosis

1. Isolation and identification of virus from blood.
2. Serological test like Haemagglutination-inhibition test.

Disease Management in Animals

1. Symptomatic treatment like supportive therapy to manage hemorrhages and dehydration.
2. Vaccination with killed KFD vaccine.
3. Control of tick vectors.

Disease Management in Humans

1. Symptomatic treatment of infected persons.
2. Vaccination with killed KFD vaccine.
3. Control of ticks by power equipment or aircraft mounted equipment to dispense carbaryl, fenthion or propoxur at 2.24 Kg of active ingredient per hectare. The spraying must be carried out in "hot spots" that is in the areas where monkey deaths have been reported within 50 m around the spot of the monkey deaths, besides the endemic foci.
4. Personal protection of individuals exposed to the risk of infection by adequate clothing and insect repellents (DMP: dimethyl phthalate) should be encouraged.
5. Health education of the public involves the habit of sitting or lying down on the ground in endemic areas should be discouraged.

Chapter 34

Zoonoses Caused by Bunyaviruses

Bunyaviruses cause important zoonoses. The genus *Bunyavirus* belong to the family Bunyaviridae. The Bunyaviridae is divided into **ar**thropod-**bo**rne viruses (**arbo**viruses) and **ro**dent-**bo**rne viruses (**robo**viruses). They are spherical, enveloped RNA viruses. On the basis of structure, genetics and ecology, bunyaviruses are grouped into four genera such as *Bunyavirus*, *Hantavirus*, *Nairovirus* and *Phlebovirus*. Bunyaviruses are responsible for a number of febrile diseases in animals and humans. The distribution of each disease caused by them is determined by the distributions of the vector and vertebrate host. Except for Hantaviruses, biologic transmission occurs by a tick, mosquito, midge, or sandfly vector. If arthropods once infected, they remain infected for life. Transovarian transmission is common in arthropods. Domestic or wild vertebrates usually are needed to maintain the cycle of infection. Humans are usually dead-end hosts for all these viruses except Phleboviruses. There may be a secondary nosocomial spread of Crimean-Congo hemorrhagic fever. Hantaan virus cycles among rodents are maintained, probably by aerosol or fomites transmission from infected rodent urine. Human infection is incidental. Bunyaviruses cause fevers sometimes with rash. In addition, they may cause hemorrhage (*e.g.,* Crimean-Congo hemorrhagic fever), hemorrhage, renal failure and pulmonary syndrome (*e.g.,* Hantaviruses infections), encephalitis (*e.g.,* La Crosse virus and related viruses) and hemorrhagic hepatitis, encephalitis, or blindness (*e.g.,* Rift Valley fever).

Crimean-Congo Hemorrhagic Fever (CCHF)

Crimean-Congo hemorrhagic fever (CCHF) is an acute viral zoonotic disease caused by *Nairovirus* (an arbovirus) and transmitted by tick. CCHF virus causes severe viral hemorrhagic fever outbreaks. The disease was first described in 1944 on the Crimean peninsula. Later the etiological agent was confirmed in 1945.

Etiology

Crimean-Congo hemorrhagic fever is caused by *Nairovirus* which belongs to the family Bunyaviridae. There are three subtypes of CCHF virus.

Epidemiology

The countries from where the CCHF has been reported are Bulgaria, Albania, Serbia, the former USSR, the UAE, Oman, Iran, Iraq, Kuwait, Afghanistan, Pakistan, China, DRC, Sierra Leone, South Africa, Australia *etc.* In the former USSR, the occurrence of CCHF is seasonal. In Iraq, CCHF occurs sporadically throughout year. CCHF is endemic in Africa, the Balkans, the Middle East and some Asian countries. In endemic areas, CCHF virus can be found in domestic animals and humans. The disease mainly occurs on livestock farms, ostrich farms, slaughterhouses, rendering plants *etc.* indicates the involvement of occupational groups. The disease is more frequently found in cattle than in the sheep or goats. Birds are also infected with this virus. The infection is more frequently found in ostriches than the other birds. CCHF was introduced by camels into the United Arab Emirates from Afghanistan. The important reservoirs of this virus are hedgehogs, horses and mouse-like rodents. The virus has also been isolated from more than 30 species of ticks. The virus is predominantly found in *Hyalomma* spp. of tick. This tick spp. is most efficient vector for transmission of CCHF virus. However, the virus is also reported from *Ixodes* spp. The distribution of CCHF coincides with the occurrence of *Hyalomma* ticks.

Transmission

1. **Tick bites:** The disease is transmitted by ticks particularly *Hyalomma* spp. Human is generally infected due to transmission of virus by infected ticks from livestock.

2. **Contact:** Activities like slaughter, castration, branding of animals *etc.* favour the transmission of disease. Unhygienic conditions favour the transmission of disease in patient caretakers. Human-to-human transmission can occur, resulting from close contact with the blood, secretions, organs or other bodily fluids of infected persons.

3. **Nosocomial infections:** Hospital-acquired infections can also occur due to improper sterilization of medical equipment, reuse of needles and contamination of medical supplies.

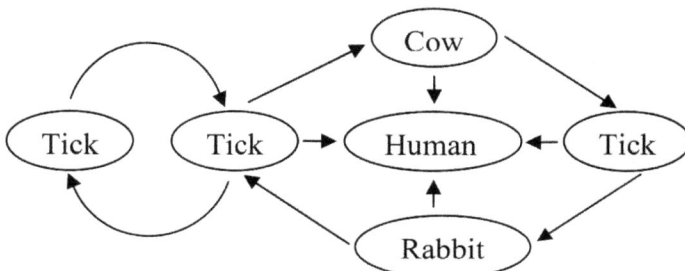

Figure 34.1: Infection Cycle of CCHF.

Disease in Animals

The hosts of the CCHF virus include a wide range of wild and domestic animals such as cattle, sheep and goats. Many birds are resistant to infection, but ostriches are susceptible and may show a high prevalence of infection in endemic areas. Animals become infected due to infected tick bites and the virus remains in their bloodstream for about one week after infection, allowing the tick-animal-tick cycle to continue when another tick bites. Although a number of tick genera are capable of becoming infected with CCHF virus, but *Hyalomma* spp. is the principal vector.

Disease in Humans

The length of the incubation period depends on the mode of acquisition of the virus. The incubation period of CCHF virus is 1 to 3 days following infected tick bite. However, incubation period may be up to 9 days in case of nosocomial infection acquired through exposure to viremic animal blood, tissues and excreta. The disease is characterized by sudden onset of fever, shivering, malaise, irritability, headache, bodyache, depression, tachycardia, lymphadenopathy, nausea, abdominal pain and hepatomegaly. Fever usually lasts for 5 to 12 days but recurrent or biphasic courses are seen. Congestion and edema are seen in conjunctiva and mucous membranes. Hemorrhagic diathesis may be seen in most of the patients on day 4 or 5. Petechial bleedings on the skin of entire body and bleeding on mucous membranes, hematemesis, melena and urogenital bleeding are seen. The case fatality rate may be 30 to 50 per cent. Death occurs usually due hemorrhagic shock or from secondary infections.

Diagnosis

Tests on patient samples present an extreme biohazard risk and should only be conducted under maximum biological containment conditions (BSL-4 laboratories). However, if samples have been inactivated (*e.g.*, with virucides, gamma rays, formaldehyde, heat, *etc.*), they can be manipulated in a basic biosafety environment. The disease can be diagnosed on the following basis-

1. **Information on tick bite:** Information on tick bite and report on agricultural work in areas of endemicity may give clue about disease.
2. **Isolation of the virus:** Virus can be isolated in Vero cells from blood of severely affected patients.
3. **Serological examination:** ELISA is important test for diagnosis of the disease. The virus can also be demonstrated in Vero cells or in the brain of intracerebrally injected mice using immunofluorescence technique.
4. **Molecular technique:** Patients with fatal disease, as well as in patients in the first few days of illness, a measurable antibody response usually do not develop. Therefore, diagnosis in these individuals can be achieved by virus or RNA detection in blood or tissue samples. A direct RT-PCR (two-step nested PCR) can be performed on patient's blood for detection of the virus.

Differential Diagnosis

It should be made with the patients having manifestations of hemorrhagic fever such as typhus, borreliosis, yellow fever, Kyasanur forest disease, dengue fever, Omsk hemorrhagic fever, malaria, leptospirosis and infections with *Hantavirus*, *Ebola* virus and *Marburg* virus.

Disease Management in Humans

1. **Use of acaricides:** It is difficult to prevent or control CCHF infection in animals and ticks as the tick-animal-tick cycle usually goes unnoticed and the infection in domestic animals is usually not apparent. Furthermore, the tick vectors are numerous and widespread, so tick control with acaricides is only a realistic option for well-managed livestock production facilities.

2. **Quarantine measures:** Quarantine measures have been effective in controlling CCHF. During an outbreak at an ostrich abattoir in South Africa, measures were taken to ensure that ostriches remained tick free for 14 days in a quarantine station before slaughter. This decreased the risk for the animal to be infected during its slaughtering and prevented human infection for those in contact with the livestock.

3. **Protective measures:** Use of gloves in contact with patients suspected of suffering from viral hemorrhagic fever and contact with potentially infected cattle, sheep, goats and camels is advised. Avoid close physical contact with CCHF-infected people, wear the gloves, use protective equipment when taking care of patients, wash the hands regularly after caring for or visiting diseased persons, follow the safe injection practices and safe burial practices. Samples taken from persons with suspected CCHF should be handled by trained staff working in well-equipped laboratories.

4. **Health education:** To reduce the risk, wear the protective and light coloured clothing for easy detection of ticks on the clothes. Use of approved acaricides on clothing, approved repellent on the skin and clothing, regular examination of clothing and skin for ticks if found, remove them safely, seek to eliminate or control tick infestations on animals or in animal houses, and avoid areas where ticks are abundant and seasons when they are most active are also important to prevent the spread of disease. Wearing gloves and other protective clothing while handling animals or their tissues in endemic areas, especially during slaughtering can reduce the risk of animal-to-human transmission.

5. **Vaccination:** An inactivated mouse brain vaccine has been developed by Russia. However, vaccination is less commonly employed due to rare occurrence of disease. There is no vaccine available for either humans or animals.

6. **Treatment:** Symptomatic and supportive treatment is the main approach to manage CCHF. RBCs, platelets, clotting factor and albumin are required for treatment of hemorrhagic shock. Antiviral drug like ribavirin 4g daily

initially and followed by 16 mg/kg body weight i.v., 4 times a day is useful.

Rift Valley Fever (RVF)

Rift Valley fever (RVF) is a viral zoonotic disease that primarily affects animals but also has the capacity to infect humans. Infection can cause severe disease in both animals and humans. The disease also results in significant economic losses due to death and abortion among RVF-infected livestock.

Etiology

The disease is caused by RVF virus, a member of the genus *Phlebovirus* in the family *Bunyaviridae*. It was first reported in livestock in Rift Valley, Kenya.

Epidemiology

RVF is generally found in regions of eastern and southern Africa where sheep and cattle are raised, but the virus exists in most of sub-Saharan Africa, including West Africa and Madagascar. Environmental factors, particularly rainfall, seem to be an important risk factor for outbreaks. Epizootic events and outbreaks in humans have been observed during years in which unusually heavy rainfall and localized flooding occur. Occupational groups such as herders, farmers, slaughterhouse workers, and veterinarians are at higher risk of infection. In epizootic events, there is increased handling of infected animals and thereby increases risk of exposure for humans.

RVF virus was first identified in 1931 during an investigation into an epidemic among sheep on a farm in the Rift Valley, Kenya. Since then, outbreaks have been reported in sub-Saharan Africa. The most notable RVF epizootic occurred in Kenya in 1950-1951, resulting in the death of an estimated 100,000 sheep. In 1977 an explosive outbreak was reported in animals and humans in Egypt, the RVF virus was introduced to Egypt via infected livestock trade along the Nile irrigation system and caused over 600 human deaths. In 1997-98, a major outbreak occurred in Kenya, Somalia and Tanzania following El Nino event and extensive flooding. Following infected livestock trade from Africa, RVF spread in September 2000 to Saudi Arabia and Yemen, marking the first reported occurrence of the disease outside the African continent and raising concerns that it could extend to other parts of Asia and Europe. The disease outbreak has also been reported in recent years in 2003, Egypt; 2006, Kenya, Somalia and Tanzania; 2007, Sudan; 2008-9, Madagascar; 2010, Republic of South Africa; 2012 Republic of Mauritania; and 2016, Republic of Niger.

Transmission

In humans, the disease is mainly transmitted due to direct or indirect contact with the blood or organs of infected animals. In humans, the disease is transmitted through handling of animal tissue during slaughtering or butchering, assisting with animal births, conducting veterinary procedures, or from the disposal of carcasses or fetuses. Transmission of human infections may occur via wound through contaminated knife or through contact with broken skin. Infections may also be transmitted through inhalation of aerosols produced during the slaughter of infected animals. There is some evidence that humans may become infected with RVF virus by ingesting the unpasteurized or inadequately heat treated milk of infected animals. Humans can be infected with RVF virus from infected mosquitoes (*Aedes* and *Culex*) bites and, rarely, from other biting insects that have virus-contaminated mouthparts. Human-to-human transmission of RVF has not been documented.

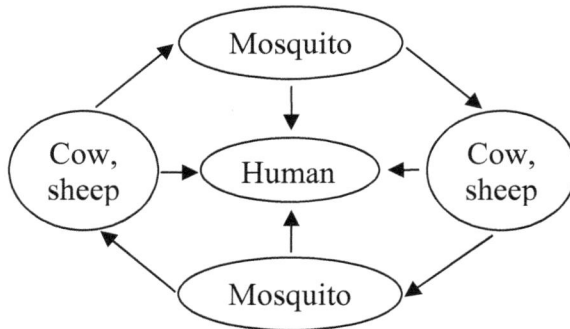

Figure 34.2: Infection Cycle of RVF.

Disease in Animals

RVF virus causes severe disease in domesticated animals including cattle, sheep, goats and camels. Sheep and goats are more susceptible than cattle or camels. Young age group of animals is more susceptible. The case fatality rate in lambs is more than 90 per cent, while, in adult sheep approximately 10 per cent. The disease causes almost 100 per cent abortions among pregnant infected ewes.

Disease in Humans

The incubation period of RVF virus varies from 2 to 6 days. The disease may occur in mild or severe form. The mild form of the disease is characterized by sudden onset of fever, headache, myalgia and arthralgia. In some patients, there may be neck stiffness, sensitivity to light, loss of appetite and vomiting. The severe form of RVF is characterized by ocular problems, meningoencephalitis or haemorrhagic fever. Ocular form of the disease is characterized by the usual symptoms associated with the mild form of the disease accompanied by retinal lesions and impaired visions. The case fatality rate of this form is less than 1 per cent. Meningoencephalitis form of the disease is characterized by intense headache, loss of memory, hallucinations, confusion, disorientation, vertigo, convulsions, lethargy and coma. Neurological complications can appear later. The case fatality rate is low (less than 1 per cent). Haemorrhagic fever form of the disease is characterized by jaundice, vomiting blood, passing blood in the faeces, a purpuric rash or echymoses, bleeding from the nose or gums, menorrhagia and bleeding from venepuncture sites. In this form of disease, the case-fatality rate is approximately 50 per cent.

Diagnosis

1. **Isolation of the virus:** The virus can be isolated from blood of early phase of illness and post-mortem tissue samples using cell culture.

2. **Clinical examination:** Clinical diagnosis of RVF is often difficult because of varying and non-specific symptoms, especially during early phase of disease. RVF is difficult to distinguish from other viral haemorrhagic fevers as well as many other diseases that cause fever, including malaria, shigellosis, typhoid fever, and yellow fever.

3. **Serological examination:** Confirmatory diagnosis can be made by detection of IgG and IgM antibodies, using ELISA.

4. **Molecular examination:** Confirmatory diagnosis can also be made by using reverse transcriptase polymerase chain reaction (RT-PCR) assay.

Disease Management in Animals

1. **Vaccination:** The outbreaks of RVF in animals can be prevented by a sustained programme of animal vaccination. Both modified live attenuated virus and inactivated virus vaccines have been developed for use in animals. The live attenuated virus vaccine requires only single dose to provide long-term immunity but this vaccine may result in spontaneous abortion in pregnant animals. The inactivated virus vaccine requires multiple doses to provide protection, but does not cause abortion in pregnant animals.

2. **Restriction on movement:** Restriction or ban on the movement of livestock may be effective in slowing the expansion of the virus from infected to uninfected areas.

3. **Surveillance:** As outbreaks of RVF in animals precede human cases, the establishment of an active animal health surveillance system to detect new cases is essential in providing early warning for animal heath and public health authorities.

Disease Management in Humans

1. **Vector control:** It is very important to control the vectors (mainly mosquitoes) of RVF virus to prevent the spread of the disease. Mosquito population can be controlled by using insecticides, larvivorous fish and other methods. Persons can protect themselves by using mosquito repellents and bed-nets.

2. **Public health education:** During an outbreak of RVF, close contact with animals, particularly with their body fluids, either directly or via aerosols, has been identified as the most significant risk factor for RVF virus infection. Raising awareness of the risk factors of RVF infection as well as the protective measures in individuals can be taken to prevent mosquito bites is the only way to reduce human infection and deaths. Public health messages for risk reduction should also focus on reducing the risk of

animal-to-human transmission arising from the unsafe consumption of fresh blood, raw milk or animal tissue. In the epizootic regions, all animal products such as blood, meat, milk *etc.* should be thoroughly cooked before consumption. Public health messages for risk reduction should also focus on the importance of personal and community protection against mosquito bites through the use of mosquito nets, personal insect repellent, light coloured clothing (long-sleeved shirts and trousers) and avoid outdoor activity at peak biting times of mosquitoes.

3. **Surveillance:** Surveillance is important to formulate effective measures for reducing the number of infections.

4. **RVF forecasting and climatic models:** Forecasting can predict climatic conditions that are frequently associated with an increased risk of outbreaks, and may improve disease control. In Africa, Saudi Arabia and Yemen, RVF outbreaks are closely associated with periods of above-average rainfall. The response of vegetation to increased levels of rainfall can be easily measured and monitored by Remote Sensing Satellite Imagery. In addition RVF outbreaks in East Africa are closely associated with the heavy rainfall that occurs during the warm phase of the El Nino–Southern Oscillation (ENSO) phenomenon.

5. **Infection control in health care settings:** Although no human-to-human transmission of RVF has been demonstrated, there is still a theoretical risk of transmission of infection from infected patients to healthcare workers through contact with infected blood or tissues. Healthcare workers caring for patients with suspected or confirmed RVF should implement standard precautions when handling specimens from patients. Standard precautions include handling of blood, other body fluids, secretions and excretions.

6. **Vaccination:** An inactivated vaccine has been developed for human use. However, this vaccine is not licensed and is not commercially available.

7. **Treatment:** Mild form of disease requires no specific treatment. Severe form of disease requires symptomatic and supportive treatment.

Hantavirus Infections (*Hantavirus* Hemorrhagic Fever with Renal Syndrome and *Hantavirus* Pulmonary Syndrome)

Synonyms

Four corners disease, Sin Nombre disease, Muerto Canyon virus disease

Hantavirus infections are emerging viral zoonotic diseases that are spread to humans through rodents. Unlike other Bunyaviridae viruses, Hantaviruses are transmitted through inhalation of the virus from infected rodent faeces. *Hantavirus* infections may lead to *Hantavirus* hemorrhagic fever with renal syndrome (Old world *Hantavirus*) and pulmonary syndrome (New world *Hantavirus*). *Hantavirus* pulmonary syndrome became known in May, 1993 as a new *Hantavirus* disease in the US.

Etiology

The causative agent belongs to the genus *Hantavirus,* a member of the family Bunyaviridae. Small rodents such as rats, voles *etc.* serve as reservoirs of different Hantaviruses. *Hantavirus* together with Arenaviruses is also called Roboviruses (rodent-borne viruses) for ecological reason. The genus *Hantavirus* is exceptional in the Bunyaviridae viruses, because arthropod vectors are not required in the transmission cycle of *Hantavirus.*

Epidemiology

Hantaviruses are the most widely distributed zoonotic rodent-borne viruses, and can be found in North and South America, Europe, and the Asia-Pacific region, including India, Indonesia, Myanmar, Sri Lanka and Thailand. *Hantavirus* infections occur mostly in the summer dry season.

Transmission

The source of human infection with *Hantavirus* is water contaminated with mouse excrements or aerosols. The infection may be transmitted through the following routes:

1. **Ingestion:** In humans, the disease may be transmitted due to consumption of food contaminated with rodent excrements. Human-to-human transmission of infection has been reported.

2. **Inhalation:** The disease can also be transmitted due to inhalation of dusts contaminated through farm work, mining and boot camps in rodent infected areas.

3. **Contact:** Persons working or staying for long periods in room that are infected with rodents and their excrements.

4. **Nosocomial infection:** Nosocomial infection is possible but occurrence is rare.

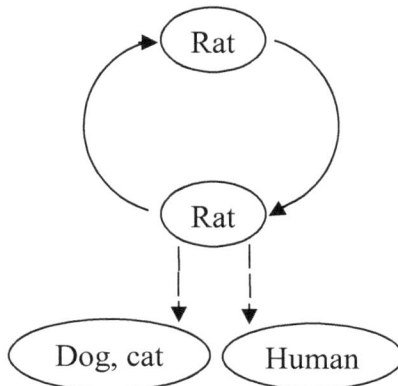

Figure 34.3: Infection Cycle of *Hantavirus.*

Disease in Humans

The disease is of two types:

1. ***Hantavirus* hemorrhagic fever with renal syndrome (HFRS):** It affects the kidney and is more common in Asia. The disease is characterized by sudden onset of high fever, chills, and pain in the back, muscles, and abdomen. Profuse bleeding may occur from haemorrhage, and kidney function can also be impaired. Tiny red spots and rashes appear on the face and elsewhere. Bleeding may also be seen from the nose and in the urine. There may be sudden drop in blood pressure and leads to shock. In severe cases there may be involvement of nervous system which may cause encephalitis or Guillian-Barre syndrome. The case fatality rate may be up to 15 per cent.

2. ***Hantavirus* pulmonary syndrome (HPS):** The incubation period is usually 2-3 weeks for the renal syndrome. The most common route of transmission is inhalation of contaminated aerosols in house infested with deer mice. It affects the lungs and is more commonly seen in the Americas. An initial period of general fever, ache and cough is followed by the sudden onset of acute respiratory distress. Nausea, vomiting, diarrhea, abdominal pain, headache rigor, malaise, myalgia and arthralgia also occur in the early stages. Other symptoms include respiratory distress, sweating, back pain, daze, thrombocytopenia, generalized edema and pulmonary edema. Death occurs mainly due respiratory failure. The case fatality rate may be up to 50 per cent.

Disease in Animals

Persistent infection and virus excretion without clinical signs has been observed in mice and rats (serve as reservoirs of *Hantavirus*).

Diagnosis

1. **Clinical examination:** HFRS can be tentatively diagnosed by symptoms of fever, increasing renal insufficiency, edema and hemorrhagic diathesis. The HPS shows high fever, increase in lung density and respiratory insufficiency.

2. **Epidemiological surveillance:** Epidemiological situation and possible exposure to rodents should be taken into consideration.

3. **Serological diagnosis:** Neutralizing antibodies are present in serum at the onset of acute disease, which is in contrast to hemorrhagic fever caused by *Arenavirus, Flavivirus* and *Filovirus*, where neutralizing antibodies are only found during the recovery phase.

4. **Molecular diagnosis:** *Hantavirus* can be detected by RT-PCR.

Disease Management in Humans

1. **Personal hygiene:** It is important in infected areas. Patients and suspected patients can be kept in negative pressure unit to avoid nosocomial spread

of *Hantavirus*. Laboratory personnel should be careful while handling the infected specimen or centrifuging it to avoid aerosol transmission of infection.

2. **Antiviral therapy:** Ribavirin (@4 g per day initially followed by 16 mg per kg body weight, i.v., six hourly) can be effective for treatment of the disease.

3. **Symptomatic treatment:** It includes treatment that supports the circulation and renal functions.

4. **Other measures:** Avoid contact with wild rodents. Persons cleaning rodent-infested structures in endemic areas should wear protective clothing and avoid inhaling or creating dust/aerosols. Eliminate the rodents from homes and prevent them from re-entering.

Chapter 35

Zoonoses Caused by Filoviruses

Filoviruses belong to the family Filoviridae and can cause severe hemorrhagic fever in humans and nonhuman primates. Filovirus virions (complete viral particles) exhibit polymorphism that is may appear in several shapes like branched filamentous, shorter filamentous-shaped like a "6", a "U" or a circle. Each virion contains one molecule of single-stranded, negative-sense RNA. The family Filoviridae includes three genera, namely, Cuevavirus, Marburgvirus, and Ebolavirus. There are five species of *Ebola* virus that have been identified as Tai Forest (formerly Ivory Coast), Sudan, Zaire, Reston and Bundibugyo. The three *Ebola* virus species namely, Bundibugyo, Zaire and Sudan have been associated with large outbreaks in Africa. The virus causing the 2014 West African outbreak belongs to the Zaire species. *Ebola* Reston is the only known filovirus that does not cause severe disease in humans, however, it can still be fatal in monkeys and it has been recently isolated from infected swine in Southeast Asia.

Ebola Virus Hemorrhagic Fever

Ebola virus hemorrhagic fever is a highly fatal viral zoonotic disease. All known epidemics have occurred as a result of nosocomial infections with unknown primary infections. The role of arthropod vectors in transmission of *Ebola* virus has not been found. The epidemics of *Ebola* virus have occurred predominantly at the end of the rainy season.

Etiology

The disease is caused by *Ebola* virus which belongs to the family Filoviridae. The two important genera of Filoviridae are *Ebola* virus and *Marburg* virus. They are enveloped, single-stranded negative-sense RNA virions.

Epidemiology

There are four subtypes of *Ebola* virus. Three distinct subtypes of *Ebola* virus were isolated from Africa. The strain Maridi was first time isolated from southern Sudan in 1976. Later on this strain was isolated from Gulu in 1979 and north-west Uganda in 2000. In these outbreaks the fatality rate has ranged from 32.5-65 per cent. The Zaire subtype of *Ebola* virus originally isolated from a patient in Yambuku. Later on this subtype was reappeared in Kikwit in 1995. This subtype was also isolated from patients in several outbreaks in Gabon in the Democratic Republic of the Congo (formerly Zaire) and more recently from outbreaks in the Republic of Congo. In these outbreaks the fatality rates has ranged from 60-90 per cent. The third subtype Ivory Coast was isolated only in association with an outbreak in chimpanzees in the Tai forest in Ivory Coast. The fatality rate was high in chimpanzees. The fourth subtype Reston virus has been isolated from cynomolgus monkey in Reston during 1989-1990 and in Italy in 1991. This subtype has also been isolated in Philippines from monkeys. It was the first subtype of filovirus virus from outside of Africa.

An outbreak of *Ebola* virus hemorrhagic fever occurred during 2000 in Gulu, Uganda. Total 329 cases were reported in which the fatality rate was 32.5 per cent. *Ebola* virus disease occurred in Democratic Republic of the Congo and Gabon in December 2001. Total 32 cases were reported with 72 per cent fatality rate. In these outbreaks the important sources of *Ebola* virus were monkey, chimpanzee and humans. However, the transmission cycle and reservoir of filovirus have remained unknown. Monkeys were the source of human infection with Reston virus from the Philippines. In Africa, human infection with the subtype Zaire and Ivory Coast were associated with fatal disease in chimpanzees. Monkeys are the only known sources of filovirus infection acquired by humans. However, they can hardly be the reservoir of virus because of the high pathogenicity of filovirus in monkeys.

Transmission

1. **Contact:** Disease transmission in cases has been found due to personal contact or contact with blood. Disease transmission was also found due to bare hand contacts with deceased during burial rituals in Africa.

2. **Inoculation:** Disease transmission in cases has also been found due to accidental inoculation. In the Zaire epidemic during 1976, half of the cases were associated with transmission of disease due to inoculation with reused needles in the hospital.

3. **Ingestion:** Transmission of disease has also been reported due to ingestion of meat from infected chimpanzees. In Gabon, the disease transmission has been occurred on several occasions due to consumption of dead chimpanzees as food. This route of transmission has also been reported in recent epidemics in Republic of Congo.

Disease in Humans

The incubation period of disease is 6-9 days but may vary from 2 to 21 days. The disease is characterized by sudden onset of fever, severe malaise, headache, myalgia, pain in the neck and abdomen and diarrhea. Hemorrhagic diathesis has been found in 75 per cent cases, while maculopapular exanthema in 50 per cent cases. The fatality rate varies according to the pathogenicity caused by different strains of the *Ebola* virus. Reston virus is not pathogenic for humans.

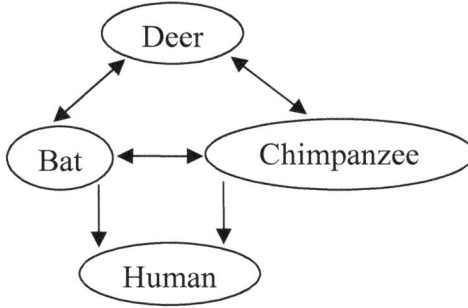

Figure 35.1: Infection Cycle of *Ebola* Virus Hemorrhagic Fever.

Diagnosis

The specimens of plasma, EDTA-treated blood, urine or liver and spleen from deceased persons are suitable for detection of *Ebola* virus by following methods:

1. **Microscopic examination:** Electron microscopy is useful for detection of *Ebola* virus.

2. **Serological examination:** *Ebola* virus can be detected using antigen capture ELISA technique. A serological diagnosis based on IgM capture ELISA is superior to immunofluorescence towards the end of the first week of the disease. ELISA technique is used in epidemiological surveys.

3. **Isolation:** *Ebola* virus can be isolated using cell culture (Vero cell clone E6).

4. **RT-PCR technique:** There are considerable variations in the sequence of different African strains of *Ebola* virus. RT-PCR has been found useful for the diagnosis in the suspected cases.

Disease Management in Humans

1. **Prevention of virus transmission:** It is important to prevent the virus transmission that may occur due direct contact during cremation of the person died of *Ebola* virus infection. Safe handling of infected persons as well as infected materials is very important to prevent the spread of *Ebola* virus. Experimental work should be performed in biosafety level 4 (BSL-4) laboratories. Specimens must be inactivated by fixation or by thermal (56°C for 30 minutes) or chemical treatment before removing from the isolation ward. Special precautions should be taken while handling or subjecting the specimen to centrifugation.

2. **Vaccination:** Vaccines have been tested in the animals. The best protection has been found with a GP DNA vaccine followed by a booster inoculation of a recombinant GP.

3. **Treatment:** The patients are symptomatically treated to control the hemorrhagic diathesis, shock *etc*. Fresh-blood transfusions are recommended to control the hemorrhagic diathesis under African conditions. Exchange transfusion should be strictly avoided. Experimentally, filoviruses are susceptible to ribavirin and interferon. Traumatic injury should be avoided that may result into hemorrhage.

Marburg Virus Hemorrhagic Fever

Marburg virus hemorrhagic fever is a highly fatal viral zoonotic disease. *Marburg* virus causes hemorrhagic fever in human beings.

Etiology

The disease is caused by *Marburg* virus which belongs to the family Filoviridae. The causative agent has been named *Marburg* virus after the place where the virus was isolated and characterized.

Epidemiology

The virus was first time isolated in 1967 from persons developed hemorrhagic fever in Marburg, Belgrade, Yugoslavia, Germany and Frankfurt. Laboratory personnel working in vaccine production were also affected. Monkeys of the species *Cercopithecus aethiops* imported from Europe to Uganda were the common sources of infection. A large number of cases were reported in 1999 and 2000 in Republic of Congo. The case fatality rate was 50 per cent.

Transmission

The disease is transmitted by contact with infected blood or organs of moneys. Aerosol transmission is also possible. Nosocomial infection may occur due to accidental inoculations or with patient blood through skin lesions.

Disease in Humans

The average incubation period is one week. The disease is characterized by sudden onset of high fever, malaise, conjunctivitis, pharyngitis, maculopapular rash, headache, nausea, vomiting, abdominal pain and diarrhea. There may be involvement of nervous system which is characterized by depression and encephalitis at the end of the acute phase.

Diagnosis

1. **Virus isolation:** The disease can be diagnosed by isolation of the virus by inoculating guinea pig or Vero cell with blood or urine.

2. **Electron microscopy:** Virus can be demonstrated by electron microscopy in formaldehyde-fixed blood after differential centrifugation.

3. **Serological method:** Viral antigen can be detected by immunofluorescence in blood smear or imprints from the post-mortem materials. Virus specific IgM antibodies can be demonstrated by IgM capture ELISA.

4. **Molecular method:** Virus can also be detected directly in EDTA-treated blood by using RT-PCR technique.

Disease Management in Humans

1. **Prophylaxis:** Extreme cautions should be taken while handling potentially infected materials from cases or monkeys. Diagnostic or experimental procedures should be performed in BSL-4 laboratories.

2. **Treatment:** Patients are treated symptomatically such as control of homeostasis and hemorrhagic diathesis.

Chapter 36

Foodborne Viral Diseases

Viruses cause many diseases in plants, animals, and humans. They are strict intracellular parasites with cellular specificity. Viral particles can be transmitted by different routes, such as contaminated food and water. People usually get infected orally, after ingestion of products contaminated during processing or subsequent handling or preparation. There are several diseases such as hepatitis E, severe acute respiratory syndrome, Nipah virus infection, monkey pox and Kyasanur forest disease which are considered as foodborne viral zoonotic diseases. However, many foodborne viruses are found in humans as well as in animals; but they generally circulate through human-to-human cycle. These viruses are not considered as zoonotic because their natural transmission does not require vertebrate animals. These viruses are transmitted to humans mainly due to poor hygiene. Norwalk-like agents are the most important causes of foodborne viral infections with 23 million cases annually worldwide followed by rotaviruses, astroviruses and hepatitis A virus. Some important foodborne diseases causing viruses are discussed as follows:

Adenoviruses

Adenoviruses are members of the genus *Mastadenovirus* which belong to the family Adenoviridae. Adenoviruses can spread not only via droplet infection, but also via the faecal-oral route. They cause 10 per cent of gastroenteritis in children and are the second most common cause of hospitalization due to diarrhea of children in Japan. Together with adenoviruses causing respiratory diseases, they can be isolated from faeces of infected children. They can be detected in sewage, sea water and shellfish.

Astroviruses

Astroviruses are the members of the genus *Astrovirus* which belong to the family Astroviridae. Human *Astrovirus* is a significant cause of acute diarrhea among children, resulting in outbreaks of diarrhea and occasionally in hospitalization. *Astrovirus* disease is generally milder than that caused by rotaviruses. However, frequent co-infection of *Astrovirus* with *Rotavirus* and *Caliciviruses* in childhood diarrhea complicates the epidemiology. Infections are more common in winter. Non-enteric symptoms like subfebrility, headache *etc.* can often be observed in grown up children on numerous occasions. The likely route of transmission is faecal-oral via food or water.

Enteroviruses

Enteroviruses are the members of the genus *Enterovirus* which belong to the family Picornaviridae. The genus *Enterovirus* is divided into five major groups such as polioviruses, group A coxsackieviruses, group B coxsackieviruses, echoviruses and newer identified enteroviruses. Designation of enteroviruses is used due to their capability to multiply in the intestine. They are quite resistant to the impact of the environment where they can survive for several weeks; they are stable in acid conditions, and consequently also in gastric juices. The human enteroviruses are ubiquitous, enterically transmitted viruses that cause a wide spectrum of illnesses among infants and children. Cases of infection giving evidence of association with eating soft fruit, green vegetables and other foods have been recognised. Enteroviruses have been frequently isolated from shellfish samples, particularly from oysters. Though enteroviruses are particularly transmitted via the faecal-oral route, the spread of certain species via aerosol is also a cause for concern. The virus can also be transmitted by flies. Viral particles are shed with faeces and symptoms of the diseases caused by them are different from typical gastroenteritis. Viruses enter the host with contaminated water or food and multiply in the digestive tract. Symptoms of infection are often slight, moderate but almost the *Enterovirus* infections are asymptomatic. However, viruses may spread into other organs and cause diseases that are serious or even fatal conditions such as aseptic meningitis, and occasionally paralysis.

Hepatitis A Virus (HAV)

Hepatitis A virus is the member of the genus *Hepatovirus* which belongs to the family Picornaviridae. It differs from enteroviruses by certain biological characteristics such as marked tropism for liver cells, exceptional thermostability (it survives heating for 30 min to 56°C), acid-resistance which can tolerates pH 1 or slow replication without cytopathic effect on the host cell. The virus is most commonly transmitted via the faecal-oral route, either by direct contact with Hepatitis A virus infected person or by ingestion of Hepatitis A virus-contaminated food or water. The virus can be transmitted via ingestion of oysters or mussels caught from seawater contaminated with human faeces. The virus can also be transmitted via fruits and salads contaminated with virus from human excretions. Washing of contaminated fresh vegetables does not guarantee elimination of viral particles.

Hepatitis E Virus (HEV)

Hepatitis E virus belongs to the family Picornaviridae. It is a major cause of outbreaks and sporadic cases of viral hepatitis in tropical and subtropical countries.

The virus is transmitted by the faecal-oral route with faecal contaminated drinking water being the usual vehicle. The virus most frequently affects young adults. The disease is usually mild, except in pregnant women, who suffer a high fatality rate from fulminant hepatic failure. The close genetic relationship of the swine and human virus suggests that swine may be a reservoir of Hepatitis E virus and swine manure could be a source of Hepatitis E virus contamination of irrigation water or coastal waters with concomitant contamination of produce or shellfish. Consumption of undercooked pig liver, and undercooked wild boar meat may cause disease. Consumption of uncooked deer meat may be a major epidemiological risk factor for Hepatitis E virus infection.

Norovirus (Norwalk-like virus)

Noroviruses are the members of the genus *Norovirus* which belong to the family Caliciviridae. *Norovirus* was formerly called Norwalk-like virus. The first recognized *Norovirus*, Norwalk virus, gained its name from an outbreak of "winter vomiting disease" in 1968 at an elementary school in Norwalk, Ohio in the USA. Noroviruses are frequently the cause of sporadic cases and also outbreaks of acute gastroenteritis in children and adults particularly in semi-closed environments such as schools, cruise ships, hospitals and residential homes. Factors that contribute to the significant impact of noroviruses include a large human reservoir, low infection dose, their environmental robustness, the short-lived immunity to noroviruses, and the ability to be transmitted by various routes. Contaminated water may be a source of infection. Viruses are present in faeces and vomitus of diseased persons. The foods most closely associated with noroviral infection are shellfish which obtain their nourishment by filtration of surrounding water and thus they ingest small particles such as seaweeds and other microorganisms including viruses, the later being concentrated within the gills of the shellfish. Outbreaks of Norwalk virus gastroenteritis following consumption of oysters have been reported in many countries. The clinical manifestation of *Norovirus* infection, however, is relatively mild. The symptoms are vomiting and diarrhea, and rarely convulsion. Asymptomatic infections are common and may contribute to the spread of the infection. Introduction of *Norovirus* in a community or population (a seeding event) may be followed by additional spread of the disease because of its highly infectious nature, resulting in a large number of secondary infections, up to 50 per cent of contacts. This virus is currently recognized as the cause of more than 96 per cent outbreaks of nonbacterial gastroenteritis in adults particularly in Australia and Europe.

Rotaviruses

Rotaviruses are the members of the genus *Rotavirus* which belong to the family Reoviridae. This genus is antigenically divided into serological groups; from A (with two to three subgroups and 11 serotypes) to E. *Rotavirus* F and *Rotavirus* G groups are provisional for the present. Groups A, B, and C of human rotaviruses have been recognized. Rotaviruses can survive for weeks in potable and recreational waters and for at least four hours on human hands. The viruses are relatively resistant to commonly used hard-surface disinfectants and hygienic hand-wash agents. Their massive excretion begins with the first day of diarrhea. They are found in waste water and can also be concentrated by shellfish; however, rotaviruses have not been linked with infectious disease following seafood consumption. Rotavirus is transmitted by faecal-oral contact and possibly by contaminated surfaces and hands and respiratory spread. Numerous animal species are infected with rotaviruses distinct from the human ones. Between human and animal rotaviruses there can be genetic reassortment though the VP6 antigen remains common in the group. Rotaviral infection may develop directly after consumption of meat from an infected animal, or indirectly by consumption of contaminated food usually eaten raw (fruit and vegetables). Human rotaviruses, particularly of group A, are considered the main cause of viral gastroenteritis in infants and young children throughout the world. Rotaviruses of group B cause gastroenteritis in adult persons. Rotaviruses of group C also cause gastroenteritis.

Prevention and control of foodborne viral diseases:

1. Strict hygienic measures must be adopted for processing, handling and storage of foods.
2. There should be systematic inspection and legislation for food safety against foodborne viral diseases.
3. The education of food industry managers, producers, distributors, and consumers about hygienic regulations and conditions of food production and processing (the use of wholesome water for watering and food processing, clean utensils *etc.*) are essential.

Fungal Zoonoses

Fungal zoonoses are caused by filamentous fungi (moulds) as well as yeasts. Moulds reproduce by the process of sporulation, while yeasts by budding. Fungal zoonoses may also be caused by dimorphic fungi (can change their growth to a mould or a yeast depending on the environmental conditions). The incidence of fungal zoonoses has increased in persons whose defense mechanisms are severely impaired by prolong and extensive use of corticosteroids, broad spectrum antibiotics, cytotoxic drugs, antineoplastic agents and concurrent diseases such as tuberculosis, diabetes mellitus, leukemia, AIDS, lymphosarcoma, cancer *etc.* On the basis of involvement of body parts, fungal zoonoses may be categorized into three groups. Superficial mycoses are the most common fungal zoonoses and affect skin, hair, nail and mucous membrane, *e.g.*, dermatophytosis infection and yeast infections particularly *Candida* spp. Subcutaneous mycoses involves skin, subcutaneous tissues and bones, *e.g.*, maduromycosis, and sporotrichosis. Systemic mycoses are deep sheeted infections involving internal organs which generally result from inhalation of air contaminated with fungi, *e.g.*, blastomycosis, coccidioidomycosis and histoplasmosis. Fungal infections can be diagnosed by direct microscopy, histopathological examination, cultural examination, animal inoculation tests, serological tests and molecular techniques. Direct demonstration of fungal pathogen in clinical materials and its isolation in pure and luxuriant form are still considered the best methods to conclusively confirm the diagnosis. Adoption of high standards of hygiene such as use of protective clothing, isolation of patients, care during handling of diseased animals or infected materials, thorough washing of hands with antiseptic solution after examination of patients *etc.* are important in prevention of fungal zoonoses. Health education to the public about source of infection and severity of disease should be imparted to prevent the spread of disease. Dermatophytosis and cutaneous candidosis can be treated with topical application of 2 per cent ketoconazole or 1 per cent clotrimazole. Oral therapy with amphotericin B 50-100 mg or ketoconazole100-200 mg is recommended in most of the fungal diseases. Systemic mycoses can be effectively treated with amphotericin B (@ 0.1-1.0 mg/kg body weight, i.v.). Despite known toxicity, amphotericin B is still considered as the gold standard for treatment of fungal zoonotic diseases. Some important fungal zoonoses are discussed in next chapters.

Chapter 37

Aspergillosis

Aspergillosis is an acute or chronic granulomatous disease of many animal species, birds and humans. Aspergillosis most frequently affects the lungs, but infection may occur at other sites such as the nasal sinuses and superficial tissues.

Etiology

The disease is caused by *Aspergillus fumigatus*, *A. flavus*, *A. nidulans* and *A. niger*. *A. fumigatus* is the most common species in causing aspergillosis, particularly invasive form.

Epidemiology

Aspergillosis occurs worldwide. In India, the disease has been reported in animals, birds and human beings. The disease has been associated with occupational groups such as farmers and other occupational workers.

Transmission

The disease is transmitted due to inhalation of air contaminated with spores of the causative agent. Spores of *Aspergillus* spp. are widely distributed in the soil of dairy and poultry farms and zoological gardens.

Disease in Animals

In domestic animals, the lesions are found in the skin, liver, lungs, spleen and lymph nodes. In cattle, the disease may cause abortion. Clinically manifested animals may show fever, cough, enlargement of lymph nodes, hemorrhages in the heart and other internal organs and nervous disorders. In birds, the disease may occur in acute or chronic form. Acute form is found mostly in young birds and characterized by fever, diarrhea, gasping and convulsion. Chronic form is mostly found in older age group of birds and characterized by mild respiratory illness with necrotic and granulomatous lesions.

Disease in Humans

In humans, *Aspergillus* spp. may cause invasive aspergillosis, paranasal granuloma, aspergilloma of lungs and endocarditis. Immune suppression, organ transplantation and excessive use of corticosteroids increase the risks of acquiring aspergillosis.

Diagnosis

1. **Histopathological examination:** Exudate, aspirate, biopsy and autopsy materials collected from infected humans and animals can be examined satisfactorily using per iodic acid Shiff's (PAS) staining technique.

2. **Cultural examination:** Clinical materials can be cultured by using Sabouraud's dextrose agar medium for confirmatory diagnosis.

3. **Serological examination:** Immunodiffusion test can be applied for diagnosis of aspergillosis.

Disease Management in Animals

1. **Hygienic measures:** High standards of hygiene should be maintained at livestock and poultry farms. Decontamination of farm buildings can be carried out by using 5 per cent formalin.

2. **Treatment:** Amphotericin B can be used for the treatment of aspergillosis.

Disease Management in Humans

1. **Hygienic measures:** Thorough washing of hands with antiseptic solution after examination of the patients should be followed.

2. **Health education:** Health education to the public about source of infection, severity of disease and personal hygiene should be imparted to prevent the spread of disease.

3. **Treatment:** Allergic aspergillosis can be treated with corticosteroids; invasive aspergillosis with amphotericin (@ 0.3mg/kg body wt., i.v., 4 times daily for months) and aspergilloma of lungs can be treated by surgical excision of affected part.

Chapter 38

Blastomycosis

Synonyms

Chicago disease, Gilchrist disease

Blastomycosis is a chronic infection of the lungs which may spread to other tissues particularly skin and bone. It occurs mainly in dogs, cats and humans.

Etiology

The disease is caused by *Blastomyces dermatitidis*.

Epidemiology

Blastomycosis occurs mainly in North America, some countries of Latin America and Africa. The disease is endemic in USA. The prevalence of disease is higher in the young male dogs. The disease has also been reported from India in birds, bats and humans.

Transmission

The disease is transmitted due to inhalation of air contaminated with spores of the causative agent.

Disease in Animals

In dogs, the disease affects mainly lungs. It localizes in the lungs but the infection may disseminate to other organs also. The clinical manifestations are fever, cough, dyspnea, skin lesions, blindness and lameness. In cats, the disease is characterized by similar manifestations as in dogs.

Disease in Humans

The disease occurs sporadically in humans. The disease is characterized by abscess and granulomatous lesions in the lungs, bone and other organs. There may be other manifestations such as fever, cough, chest pain, ulcerated lesions on the skin and lymphadenopathy.

Diagnosis

1. **Clinical examination:** In disseminated infection the chronic pulmonary disease persists and abscess and granulomatous lesions are found in most organs and body tissues including bone.

2. **Radiological examination:** Chest radiograph can resemble that of tuberculosis or carcinoma.

3. **Animal inoculation test:** Laboratory animal mainly the mice can be employed for the recovery of organism from pathological materials.

4. **Molecular technique:** DNA probe can be used for diagnosis of blastomycosis.

Disease Management in Animals

The disease in the animals can be prevented by careful handling of infected animals, avoiding accidental inoculation or contamination of open wound and safety from dog bites. The disease in dogs can be treated with amphotericin B.

Disease Management in Humans

It is essential to carefully handle the infected animals and avoid contamination of open wound to prevent the spread of disease. Disinfection can be done by using 3 per cent formalin in the endemic areas. Amphotericin B is the drug of choice. It can be given @ 0.3 mg/kg body weight, i.v., four times a day for one week.

Chapter 39

Candidosis

Synonym

Candidiasis

Superficial *Candida* infection may involve the mucous membrane of the mouth and vagina, the skin or nail.

Etiology

The disease is caused by *Candida albicans* and *C. tropicalis*. *C. albicans* accounts for 80-90 per cent of cases.

Epidemiology

The disease occurs worldwide in animals and humans. The disease may be asymptomatic or clinically affects cattle, sheep, dog, cat, swine, rodents, birds and humans. AIDS, diabetes, vitamins deficiency and excessive use of antibiotic and corticosteroids are the risk factors for development of the disease.

Transmission

The disease is transmitted by contact with infected tissues. The disease can also be transmitted through contaminated meat at slaughterhouse.

Disease in Humans

Mucosal infections are characterized by the development of discrete patches on the mucous membrane of mouth, vagina, penis *etc.*, and may become confluent and form a curd-like pseudo membrane. In oral candidosis, white fleck appears on the buccal mucosa and hard palate. *Candida* infections of the skin almost invariably occur at moist sites such as the axillae, groin, perineum and sub-mammary folds. It may cause dermatitis between finger webs.

Diagnosis

1. **Clinical examination:** The disease can be diagnosed on the basis of clinical lesions.

2. **Cultural examination:** The clinical specimens from the patients can be cultured on various mycological media such as Sabouraud's medium, dermatophyte test medium (DTM), Paul's sunflower seed medium and brain heart infusion (BHI) agar for isolation of the causative agent.

3. **Microscopic examination:** Direct microscopic examination of specimens such as skin scrapings, crusted materials and biopsies *etc.* can be examined satisfactorily in wet mounts after partial digestion of the tissues with 10-20 per cent potassium hydroxide solution. This technique is helpful to reveal the presence of fungal elements.

Disease Management in Humans

Cutaneous candidosis can be treated with topical application of 2 per cent ketoconazole. Amphotericin B @ 50mg total dose can be given orally or i.v.

Chapter 40

Coccidioidomycosis

Synonyms

Valley fever, Desert fever, California disease

Coccidioidomycosis is primarily a disease of domestic animals such as cattle, sheep, pig, horse, dog and cats and humans, which affects mainly the lungs.

Etiology

The disease is caused by *Coccidioides immitis*. It grows preferably in the soil and decaying organic matter.

Epidemiology

The main reservoirs of *C. immitis* are cattle and dogs. The disease has been reported from US and Canada. The disease is endemic in USA.

Transmission

The disease is transmitted by inhalation of air contaminated with spores of the organism. The disease may be occasionally transmitted through contamination of wound.

Disease in Animals

The disease may occur as an asymptomatic or acute disseminated infection. Asymptomatic infections are generally seen in cattle, sheep and swine. In dogs, the disease affects mainly the lungs. The infection may subsequently disseminate to liver, kidney, spleen and bones. In horses, it causes disseminated infection which is characterized by cough, respiratory distress and weight loss.

Disease in Humans

In humans, the disease may occur as asymptomatic, pulmonary disease, skin disease and disseminated infection. The disease is characterized by chronic cavitating pulmonary lesions. Other symptoms may be fever, cough, chest pain *etc.*

Diagnosis

1. **Microscopic examination:** Direct microscopic examination of exudate, discharge, blood, urine, sputum, bone marrow *etc.* can be done to reveal the presence of endospore.

2. **Cultural examination:** Culture of biopsy or clinical materials for isolation of the causative agent is useful for diagnosis of the disease.

3. **Serological examination:** Tests like immunofluorescence and haemagglutination inhibition are useful for diagnosis.

4. **Animal inoculation test:** Guinea pig can be used to inoculate specimen and demonstrate the causative agent.

5. **Molecular technique:** DNA probe can be used for diagnosis of coccidioidomycosis.

Disease Management in Animals

Strict hygienic measures including prevention of contamination of wounds, avoiding exposure to contaminated environment and careful handling of infected animals are important measures to prevent the spread of infection in the animals. The disease can be treated by using amphotericin B or ketoconazole.

Disease Management in Humans

The disease can be prevented by avoiding exposure to contaminated soil or dust. Exposure of wound should be avoided while handling infected animals. The disease particularly disseminated form can be treated with amphotericin B @ 0.3 mg/kg body weight, i.v., four times a day for one week.

Chapter 41

Cryptococcosis

Synonyms

European blastomycosis, Torulosis

Cryptococcosis occurs in acute, subacute or chronic form. The disease affects mainly the respiratory and central nervous system.

Etiology

The disease is caused by *Cryptococcus neoformans*. It is a type of true yeast and reproduces by budding. The organism is excreted in the milk of infected animals. The organism is also excreted in the urine, seminal fluid and sputum of infected person.

Epidemiology

The disease occurs in sporadic form and distributed worldwide. The important reservoirs are cattle, sheep, horse, dog, cat and birds. The organism is widely distributed in soil. Bird faeces and bat guanos enrich the soil that favours the growth of organism. Persons suffering from AIDS and cancer are at higher risk of acquiring the infection.

Transmission

The disease is transmitted by inhalation of contaminated air.

Disease in Humans

The organisms from contaminated environment reach to the lungs and localize there without causing overt pulmonary symptoms. It multiplises when person's immune system is declined and produces an active cryptococcosis. It causes small discrete nodules in the lungs and meningoencephalitis.

Diagnosis

1. **Microscopic examination:** Direct microscopic examination of sputum, pus and cerebrospinal fluid can be followed to reveal the presence of the organism.

2. **Cultural examination:** Culture of faeces, biopsy or clinical materials can be used for isolation of the causative agent.

3. **Serological examination** Immunofluorescence and latex agglutination are useful tests.

4. **Animal inoculation test:** Mice can be used to inoculate specimens and demonstrate the causative agent.

Disease Management

The disease in the animals and humans can be prevented by avoiding exposure to contaminated dust or soil. The disease can be treated with amphotericin B @ 0.3 mg/kg body weight, i.v., four times a day for one week.

Chapter 42

Dermatophytosis

Synonyms

Dermatomycosis, Ringworm

Dermatophytosis is a common disease of the stratum corneum of the skin, hair and nail. Dermatophytes invade keratin by enzymic digestion and mechanical pressure. Sometimes there is only dry scaling or hyperkeratosis but more commonly there is irritation, erythema and sometimes vesiculation. In skin infections of the body, face, and scalp, spreading annular lesions with a raised, inflammatory border are produced.

Etiology

Ringworm in humans and animals is caused by three main dermatophytes-

1. **Anthropophilic:** Those dermatophytes cause infections exclusively in human beings, *e.g., Microsporum audouinii* and *Trichophyton mentagrophytes*.
2. **Zoophilic:** These primarily occur in the animals but can also infect human beings, *e.g., M. canis, M. gallinae, T. equinum* and *T. verrucosum*.
3. **Geophilic:** Those fungi which are distributed in soil and other natural habitats, *e.g., M. gypsium*.

Epidemiology

The disease caused by *Trichophyton* spp. is worldwide. The persons at risk are farmers, veterinarians, butchers and children who play with pets. The prevalence of the disease is higher in young age group of human being. The prevalence is also high in male animals and human beings.

Transmission

Human infections are transmitted mostly due to direct contact with overtly or latently infected animals. Winter season favours the transmission of infection from animal to human beings. Indirect transmission may occur through contaminated items such as leather in riding equipment.

Disease in Animals

In sheep, the flank and neck regions are involved. In cattle, the hair becomes brittle and skin on the face appears as circular plaque-like lesions. In equines, the skin lesions are characterized by patches of alopecia. In dogs and cats, the disease is manifested as scaly lesions with crusty center on the skin.

Disease in Humans

The skin lesions are characterized by dry scaling or hyperkeratosis with erythema, irritation and sometimes vesiculation. The lesions are most commonly found on the face and scalp. Hair break-off easily and nail becomes thick and brittle.

Diagnosis

1. **Direct microscopic examination:** The specimens such as skin scrapping, crusted materials, hair, nail *etc.* can be examined satisfactorily in wet mount after partial digestion with 10-20 per cent potassium hydroxide.

2. **Cultural examination:** Clinical material can be cultured by using various mycological selective media such as Sabouraud's dextrose agar medium, dermatophyte test medium, Paul's sunflower seed medium *etc.* for isolation of the organism. Mycological media can be supplemented with chloramphenicol @ 50 mg/L of the medium to reduce the bacterial contamination. Cycloheximide can be supplemented to medium @ 500 mg/L to reduce the growth of saprophytic fungi.

Disease Management in Animals

1. **Hygienic measures:** Animals affected with dermatophytosis should not be mixed with the healthy animals to avoid the spread of infection.

2. **Treatment:** Griseofulvin has been found most effective drug for treatment of dermatophytosis in animals. Alternative drugs are clotrimazole, tioconazole, ketoconazole *etc.*

Disease Management in Humans

1. **Hygienic measures:** Persons in contact or handling the animals showing skin fungal lesions should adopt standard hygiene to prevent transmission of fungal infections.

2. **Treatment:** Dermatophytosis can be treated by using griseofulvin @ 10 mg/kg body weight daily or ketoconazole @ 200 mg daily, orally for few weeks to months. Topical spray, powder or ointment of imidazole can also be used for local application.

Chapter 43

Histoplasmosis

Synonyms

Darling's disease, Reticuloendothelial cytomycosis

Histoplasmosis is an asymptomatic or relatively mild, self-limiting pulmonary infection, although chronic or acute disseminated disease may also occur.

Etiology

Histoplasmosis is caused by *Histoplasma capsulatum*, which is an intracellular parasite found in soil enriched with the droppings of birds and bats. It is a type of dimorphic fungus.

Epidemiology

Histoplasmosis occurs worldwide. The disease usually occurs in dog, cat, bat, rodents, carnivores, humans *etc*. Infections have also been reported from horses and birds. *Histoplasma* readily found in the faeces of bird and bat due to the presence of high content of nitrogen in it. Persons who visit to caves inhabited by the bats are at greater risks of acquiring *H. capsulatum* infection. Agricultural, construction and sewer workers and sweepers are especially at a higher risk of infection. AIDS, excessive exposure to corticosteroids and organ transplants are predisposing factors for histoplasmosis.

Transmission

The disease is transmitted by inhalation of spores of the causative agent from soil contaminated with bat guano or bird droppings.

Disease in Animals

Histoplasmosis is latent in most of the domestic animals. In dogs, the disease is manifested by intermittent fever, cough, lethargy, lymphadenitis and diarrhea. In cats, the disease is characterized by the development of granulomatous nodules in the lungs and involvement of spleen and heart. In horses, it causes fever, pulmonary symptoms and abortion.

Disease in the Humans

Most cases are asymptomatic. Some infections may cause acute pulmonary histoplasmosis, characterized by high fever, headache, non-productive cough, chills, weakness, pleuritic chest pain and fatigue. Most people recover spontaneously, but in some cases dissemination can occur, in particular to the gastrointestinal tract and central nervous system. Risk of dissemination is higher in infancy, old age and severely immunocompromised individuals. The chronic form of histoplasmosis occurs mainly in adults. Large cavities develop directly from primary lesion in the lung or by reactivation of old lesions. Disseminated infection occurs most often in old age and infancy or in individuals with impaired immune responses.

Diagnosis

1. **Histoplasmin test:** This test is important in epidemiological surveys.
2. **Radiological examination:** Radiological examination may assist in clinical diagnosis of disease.
3. **Cultural examination:** Sabouraud's medium (with supplementation of antibiotic) can be used for isolation of organism.
4. **Microscopic examination:** Yeast cells can be demonstrated in freshly collected specimens of peripheral blood, sputum, pus, throat swab, tracheal washing, lung aspirate, peritoneal fluid, bone marrow, urine and biopsy materials from lymph nodes mucous membrane and other tissues. Giemsa staining of smear is advised for detection of yeast cells of *H. capsulatum* because of their small size.
5. **Serological examination:** CFT is valuable serological test for diagnosis of histoplasmosis. Direct immunofluorescence of specimens also helps in rapid and correct diagnosis.

Disease Management

1. **Treatment:** The disease can be treated with amphotericin B.
2. **Disinfection:** Disinfection of soil in endemic areas can be done by using 3 per cent formalin.
3. **Environmental sanitation:** Prevention of exposure to a contaminated niche helps in reducing the occurrence of disease in humans and animals.
4. **Miscellaneous measures:** Avoid exposure to bat-inhabited caves.

Chapter 44

Rhinosporidiosis

Rhinosporidiosis is a chronic granulomatous disease which results in the development of large polyps or wart-like lesions in the nose or conjunctiva.

Etiology

The disease is caused by *Rhinosporidium seeberi*. Humans and animals such as cattle, buffaloes, dogs and horses are most commonly affected.

Epidemiology

The disease occurs mainly in India, Sri Lanka, Argentina and Brazil. More than 80 per cent cases have been reported from India and Sri Lanka.

Transmission

The exact mode of transmission of disease is unknown. However, the disease transmission is believed to be by contaminated soil, dust and stagnant water.

Disease in Animals

The disease is characterized by development of large polyps on the nasal mucous membrane which may cause obstruction in breathing. There may be purulent nasal discharge.

Disease in Humans

The disease is characterized by the development of large polyps or wart like lesions in the nose or conjunctiva. Lesions may interfere with breathing.

Diagnosis

The disease can be diagnosed on the basis of clinical symptoms and presence of polyps in the nose and conjunctiva. Microscopic examination can reveal the presence of sporangia in the nasal exudate and affected tissues. Isolation of the organism by culture method in the laboratory is presently not available.

Disease Management in Animals

Animals should be prevented from exposure to contaminated stagnant water, soil and dust. Polyps can be surgically removed.

Disease Management in Humans

The disease can be prevented by avoiding exposure to contaminated soil or dust and stagnant water. The disease can be treated by surgical excision of the polyps.

Chapter 45

Sporotrichosis

Sporotrichosis is a subacute sporadic disease in animals and humans and characterized by papulous, occasionally ulcerating skin lesions at the site of inoculation and alongside the regional lymph channels.

Etiology

The disease is caused by *Sporothrix schenckii.* It is a dimorphic fungus. The organism is ubiquitous in nature and found in rotting woods, dead plant materials, surface water and sometimes swimming pools.

Epidemiology

The disease occurs worldwide, mainly in moist tropical and subtropical zones. The disease has been reported from India, Japan, Australia, South Africa, Latin America, USA and other countries. The disease has increased frequency of occurrence in Italy. The disease has been reported from animals like cattle, swine, equine, dog, cat, camel and fowl.

Transmission

The disease is transmitted by wound infection. The disease in persons may be transmitted by scratches or bites by cats and occasionally dogs and squirrels. The disease is also transmitted through wounds created by thorns, splinters and insect bite. Airborne transmission may occur. Person to person transmission is rare.

Disease in Humans

The incubation period is 3 to 21 days but occasionally may extend to 3 months. The disease occurs in various forms. Cutaneous form is the most common form of sporotrichosis. This form is characterized by development of small nodule at the site of inoculation and later on granulomatous lesions on the skin and subcutaneous tissues. The mucosal form of disease is characterized by development of nodules in the nose, pharynx, larynx, trachea and mouth. It may cause formation of granulomas, ulcers and local lymphadenopathy on the progression of this form of disease. Disseminated form of infection rarely occurs which leads to inflammatory conditions involving lungs, bones, joints, eyes, testes, epididymis and central nervous system.

Diagnosis

1. **Clinical examination:** The disease can be diagnosed on the basis of clinical lesions.

2. **Cultural examination:** The clinical specimens from the patients can be cultured on mycological media such as Sabouraud dextrose agar and brain heart infusion (BHI) agar for isolation of the causative agent.

3. **Microscopic examination:** Histology using periodic acid-Schiff stain is helpful in revealing the causative agent which shows round or oval cells and the typical cigar-shaped fungal elements.

4. **Serological examination:** Latex agglutination test is an important test for detection of antibodies. Precipitation and complement fixation tests can also be used for diagnosis of the disease.

Disease Management in Humans

Careful handling of infected animals, rotted wood and plant materials is important to prevent the disease. Potassium iodide is the drug of choice for the treatment of cutaneous sporotrichosis. Saturated solution of potassium iodide can be given orally with milk or juice @ 5-10 drops, three times a day and is gradually increased by 3-5 drops per day to 40-50 drops three times a day, for a period of 4-8weeks. Itraconazole @ 100-200 mg, per day, orally for 3-6 months can be used as an alternative drug. Disseminated form of infection particularly involving the central nervous system can be treated with amphotericin B @ 0.5mg per kg body weight, per day, i.v. Localized sporotrichosis like cavernous lungs can be treated surgically.

Section - V

Parasitic Zoonoses

Among zoonotic infections, parasitic pathogens affect millions of people worldwide. Parasitic zoonoses may be caused by protozoa, trematodes, cestodes, nematodes and arthropods. In recent years, parasitic zoonoses have gained importance due infections with opportunistic pathogens such as *Cryptosporidium* spp. and infections like toxoplasmosis in immunocompetent individuals. Parasitic zoonoses caused by some protozoa are transmitted by arthropod vector like mosquitoes, flies, fleas, ticks, mites, lice *etc.* Parasitic infections caused by protozoa, trematodes and cestodes may require one or more intermediate hosts for their development. The intermediate hosts may be vertebrate or invertebrate (snail or other molluscs and arthropods) and facilitates the transmission of parasites. In some parasitic infections, humans may act as final host (*e.g.*, taeniasis), intermediate host (*e.g.*, echinococcosis) and paratenic host (*e.g.*, *Toxocara* infections). Factors such as increasing international trades of foods of animal origin, increasing densities of human population and increasing trend of consumption of raw and semi-cooked foods have contributed significantly in the spread of zoonotic parasitic infections worldwide. Parasites may cause mechanical injury, toxic effects on the body, anaemia, nutritional deficiency, inflammatory reactions, mechanical obstructions of intestinal and respiratory passages and immune reactions in the hosts. Some important parasitic zoonoses are discussed in next chapters.

Chapter 46

Zoonses Caused by Protozoa

Protozoan parasites are microscopic, single-celled organisms that can be free-living or parasitic in nature. They are able to multiply in humans, which contributes to their survival and also permits serious infections to develop from just a single organism. Transmission of protozoa that live in a human's intestine to another human typically occurs through a fecal-oral route (*e.g.,* contaminated food or water or person-to-person contact). Protozoa that live in the blood or tissue of humans are transmitted to other humans by an arthropod vector (*e.g.,* through the bite of a mosquito or sandfly). Pathogenic protozoan parasites are conveniently dealt with by placing them in four groups *viz., sporozoa (e.g., Cryptosporidium parvum, Toxoplasma gondii etc.)*, amoebae (*e.g., Entamoeba hystolytica*), flagellates (*e.g., Giardia lamblia, Trypanosoma* spp., *Leishmania* spp. *etc.*) and miscellaneous group (*e.g., Babesia* spp., *Balantidium* spp. *etc.*).

Cryptosporidiosis

Cryptosporidiosis is increasingly recognized as a major cause of human diarrheal disease worldwide especially in immunocompromised patients.

Etiology

Cryptosporidiosis is caused by many species of *Cryptosporidium* but *C. parvum is* widespread in mammals. It is recognized as important pathogen in calves and lambs. The disease in humans is also caused by *C. baileyi, C. felis* and *C. meleagridis* occurring in birds, cats and turkey, respectively. Two genotypes of *C. parvum* exists that is genotype 1 (exclusively a human type) and genotype 2 (calf type, which is zoonotic and found in cattle, sheep, goat, laboratory rodents and humans).

Epidemiology

Human cryptosporidiosis occurs worldwide. The prevalence is

higher in industrialized developing countries. Malnourished children and immunocompromised individuals are at great risk of acquiring infection.

Transmission

1. **Ingestion:** The disease is transmitted by ingestion of vegetables, fruits, meat and seafood containing oocysts (can survive for months in the environment).

2. **Air-borne transmission:** The disease may be transmitted by infected aerosols predominantly in children and immunocompromised persons and causes "lethal respiratory cryptosporidiosis".

3. **Contact transmission:** Person-to-person transmission is possible by direct or indirect contact.

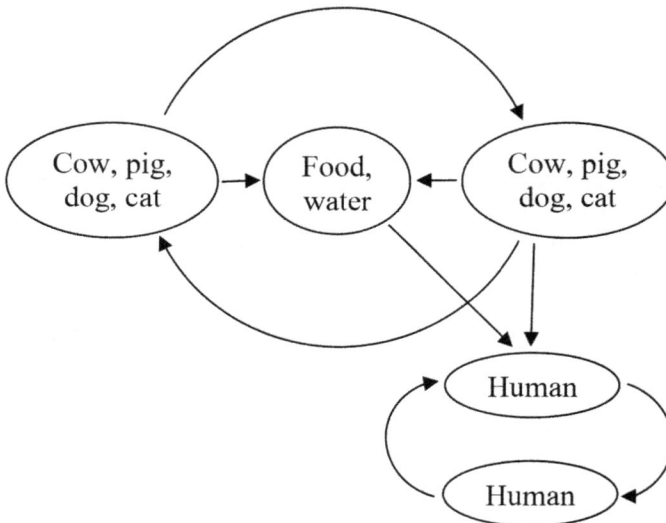

Figure 46.1: Infection Cycle of Cryptosporidiosis.

Disease in Humans

The incubation period varies from 5 to 28 days. *C. parvum* primarily- infects the intestinal epithelium but may be disseminated to epithelial cells of the other organs like bile duct or respiratory tract. The clinical manifestations are fever, nausea, profuse watery diarrhea and epigastric pain. In immunocompitent persons the infection is mostly asymptomatic.

Diagnosis

1. **Microscopic examination:** Diagnosis can be made by microscopic examination of oocysts in stool, duodenal secretions or bile.

2. **Serological examination:** Coproantigen can be detected by using ELISA technique.

3. **Staining technique:** Staining with Giemsa stain or by modified Ziehl-Neelsen technique is also useful in the diagnosis.

4. **PCR technique:** Confirmatory diagnosis can be made by using PCR technique.

Disease Management in Humans

1. **Treatment:** In immunocompitent patients, the infection is usually self-limiting and lasts for 1-2 weeks. In such patients antibiotic therapy is not required except fluid and electrolyte therapy. In immunocompromised patients, paromomycin 500 mg, orally, twice daily along with azithromycin 500 mg, orally, twice daily are recommended for 4 weeks.

2. **Hygienic measures:** Immunocompromised patients should avoid direct skin contact with human excreta or animal faeces and swimming in public baths.

3. **Miscellaneous measures:** Adequate heat treatment of foods including meat and seafood and use of filtered or boiled water may be of some measures to prevent the infection.

Toxoplasmosis

Toxoplasmosis is a systemic disease in humans and animals. The disease causes problems particularly in the children after congenital transmission and opportunistic infection in immunocompromised persons.

Etiology

Toxoplasmosis is caused by *Toxoplasma gondii*, a protozoan parasite of mammals, is transmitted when oocytes excreted by cats or present in undercooked meat are ingested. *T. gondii* belongs to the class sporozoa and subclass coccidia. Infective stages are the "sporozoites in oocyst" found in cat faeces and merozoites (cystozoites) in tissue cysts of intermediate hosts.

Epidemiology

The disease occurs worldwide and highly prevalent in humans, domestic, wild and other animals. The serological studies showed that the one third human population of the world is exposed to infection with *T. gondii*. Cat, lion and leopard are definitive host and sheds oocysts. Many worm-blooded animals such as pig, mouse and other mammals including humans and birds are intermediate host.

Transmission

1. **Ingestion:** The potential source of infection for the final host (cat) and intermediate host (pigs and man) is sporulated cysts in the environment or tissue cyst in meat. In humans, the disease is transmitted by ingestion of sporulated cysts or cysts containing raw or undercooked meat. Farm livestock acquire infection through the ingestion of oocysts from infected cats in contaminated food and water. Pigs can be infected by eating infected carrion (dead rodents, piglets *etc.*).

2. Transplacental transmission: This type of transmission is important in humans and sheep.

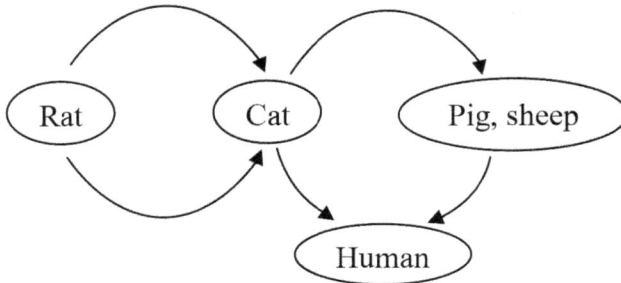

Figure 46.2: Infection Cycle of Toxoplasmosis.

Disease in Animals

In animals, the disease may be found in three forms, *viz.*, acute, subacute or chronic. The acute form of disease is characterized by diarrhea, hepatic lesions, ocular lesions, lymphadenopathy and encephalitis. In dog, cat and pig, the disease is characterized by fever, encephalitis, hepatitis, and pneumonia. Overt disease has not been reported in cattle.

Disease in Human

The severity of disease varies according to the immune status of the person. Toxoplasmosis may occur in the following forms:

1. **Acute toxoplasmosis:** In immunocompitent adult persons, 80-90 per cent infections are clinically inapparent. The disease is characterized by fever, lymphadenopathy, headache, myalgia and general weakness. The symptoms of encephalopathy, chorioretinitis, myocarditis and pneumonia are rare. In immunocompromised adult persons, more than 90 per cent of the cases appear due to reactivation of the previously acquired latent infection. Occurrence of new infections is rare. The disease is characterized by ataxia, seizures, confusion, diarrhea, dyspnea and ocular problems.

2. **Ocular toxoplasmosis:** The infection acquired in utero causes chorioretinitis. The associated symptoms are blurring of vision, scotoma, photophobia, ophthalmodynia, strabismus and loss of central vision.

3. **Pregnancy and congenital toxoplasmosis:** Immunocompitent pregnant women have relatively decreased number of CD4 helper cells (reduction of T cell mediated abortion due to immune response appears useful, since the graft rejection *i.e.* fetal abortion due to immune reaction against paternal antigens is reduced). Toxoplasmosis in pregnancy may cause abortion, still birth or prepartum delivery. Congenital toxoplasmosis in new borne may cause microcephalus, hydrocephalus, intracranial calcification, chorioretinitis, seizures, fever, anemia, jaundice, hepatomegaly, splenomegaly and lymphadenopathy. Infected new borne may survive but shows mental disorders.

Diagnosis:

Serological diagnosis like detection of specific IgM and IgG titre is important. PCR technique is important for confirmatory diagnosis.

Disease Management in Humans

1. **Hygienic measures:** Cat faeces or soil contaminated with cat faeces must be handled very carefully using disposable gloves. Vegetables and fruit should be washed carefully before their consumption.

2. **Heat treatment of meat:** Pork and lamb meat should be adequately heat treated before their consumption. Freezing at -20°C for 3 days also kills the cysts.

3. **Chemoprophylaxis:** As primary prophylaxis for AIDS patients, treatment with cotrimaxazole 750 mg twice a day, orally, is recommended.

4. **Treatment:** Pyrimethamine 25 mg per day and sulfadiazine 1 g four times, orally, daily for 3-6 weeks. Folic acid 15 mg daily orally is given to prevent bone marrow toxicity.

5. **Miscellaneous measures:** Cats should be prevented from hunting potential intermediate hosts (mice, rats *etc.*). Cats should not be fed raw meat.

African Trypanosomiasis

Synonym

Sleeping sickness

African trypanosomiasis is found mainly in tropical Africa and transmitted by tsetse flies.

Etiology

Trypanosoma brucei gambiense and T. brucei rhodesiense belong to salivaria group (transmitted via saliva), and they are important for causing disease in animals and humans. *T. brucei rhodesiense* causes more rapid and severe disease than the *T. brucei gambiense*.

Epidemiology

The disease is prevalent in Africa and Asia. *T. brucei gambiense* occurs in West and Central Africa, while *T. brucei rhodesiense* in East Africa. Human African trypanosomiasis is endemic in Sudan, Uganda, the Democratic Republic of the Congo and Angola. The disease distribution coincides with vectors. *Glossina palpalis* (tsetse fly) occurs mainly in highly humid forest zones of West and Central Africa, while, *G. morsitans* are predominant in East Africa and drier areas of savanna. Cattle and wild antelope are main reservoirs of infection.

Transmission

The disease is transmitted by tsetse flies, mainly by *G. palpalis* and *G. morsitans* bites. *T. brucei gambiense* is transmitted by *G. palpalis, G. tachinoides* and *G. fuscipes,* while *T. brucei rhodesiense* is transmitted by *G. morsitans, G. pallidipes* and *G. swynnertoni.*

Disease in Humans

The disease is characterized by development of inflammatory nodule with centrally located pustule on inoculation site (trypanosomal chancre), splenomegaly, hepatomegaly, generalized lymphadenopathy, myalgia, arthralgia, hemolysis, anemia, edema, meningoencephalitis *etc.* Lymphadenopathy is a prominent sign in patients infected with *T. brucei gambiense.* This strain causes enlargement of supraclavicular, cervical and nuchal lymph nodes.

Diagnosis

1. **Clinical examination:** The disease can be diagnosed by clinical symptoms such as fever, headache and enlargement of cervical lymph nodes.

2. **Microscopic examination:** A confirmatory diagnosis can be made by detection of trypanosomes in the fluid expressed from trypanosomal chancre in lymph node aspirate, peripheral blood, bone marrow aspirate or cerebrospinal fluid. Parasites can be detected in unstained wet mount or smear stained by the Giemsa or Leishman method. The parasite is morphologically characterized by single anterior flagellum arising via undulating membrane from a basal body situated near a posteriorly placed kinetoplast.

3. **Serological diagnosis:** ELISA, IHA and IFA are widely used for diagnosis of the disease. There is increased IgM level (3-4 times than the normal) in serum and cerebrospinal fluid.

4. **Molecular diagnosis:** PCR technique is useful for detection of trypanosomes.

Disease Management in Humans

1. **Control of tsetse flies:** Insect repellents and protective clothing can be used to control the tsetse flies and bites by them. Tsetse flies populations can also be reduced by traps and screens and by clearing the breeding places (bush vegetation near riverbanks and around settlements).

2. **Control on travel:** Avoid visiting of endemic areas of the disease and vectors.

3. **Treatment:** The choice of drugs depends on the stage of infection and *Trypanosoma* spp. involved. Infection with *T. brucei gambiense*, stage I is treated with suramin or pentamidine isethionate, and stage II with eflornithine. Suramin @ 1g per day at the days 1, 3, 7, 14 and 21, is administered i.v. A single test dose of 100-200mg is essential prior to therapy by suramin to avoid hypersensitivity reaction. Pentamidine isethionate @ 4mg per kg body weight, per day is administered i.m., for 10 days. Infection with *T. brucei rhodesiense* stage I is treated with suramin or pentamidine isethionate, and II stage by combination of melarsoprol

and suramin. Melarsoprol @ 2-3.6mg per kg body weight per day is administered i.v, three times a day for 3 days.

American Trypanosomiasis

Synonym

Chagas' disease

American trypanosomiasis is a serious public health problem in Central and South American countries.

Etiology

American trypanosomiasis is caused by a large protozoan parasite (*Trypanosoma cruzi*).

Epidemiology

About 7- 12 million people are newly infected every year in American countries with death of 45000 people from Chagas' disease. The disease is more prevalent in areas where insanitary conditions prevail. The main reservoirs of *T. cruzi* are dogs, cats, pigs, rodents, bats, opossums *etc.*

Transmission

American trypanosomiasis is transmitted by bloodsucking reduviid bugs (*Triatoma* spp.). Bugs are infected by feeding of blood (trypomastigote stage) from infected animals and humans. Trypomastigote develops into epimastigote stage in the bugs. *T. cruzi* belongs to stercoraria group (transmitted via faeces). *T. cruzi* is not transmitted by the bug bites. Human is infected from infected bug faeces (large numbers of trypanosomes are present) which is usually defecated by the bug after blood meal from the host. The unwitting sleeping person rubs the bug's faeces into the irritating bite wound.

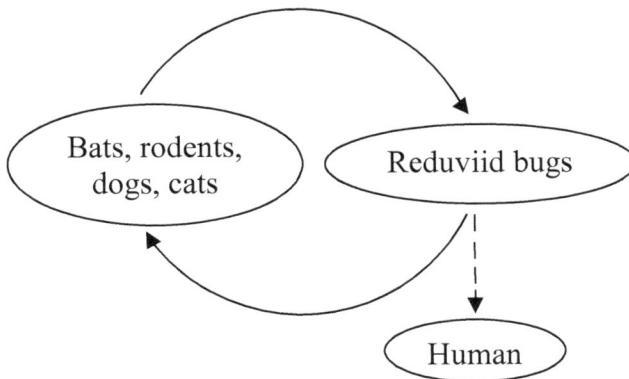

Figure 46.3: Infection Cycle of American Trypanosomiasis.

Disease in Humans

The disease is characterized by the development of erythematous swelling with pain (Chagoma) and regional lymphadenopathy at cutaneous site of entry. Other clinical manifestations are fever, general malaise, edema of face and lower extremities, hepatomegaly, splenomegaly *etc.* In some cases myocarditis and meningoencephalitis are also observed.

Diagnosis

1. **Clinical examination:** A presumptive diagnosis can be made on the basis of clinical symptoms and bug biting history.
2. **Microscopic examination:** A confirmatory diagnosis can be made by detection of trypanosomes from blood of patient in acute stage.
3. **Cultural examination:** This method is also helpful in detecting *T. cruzi* by culturing blood or other specimens. However, this method is not very useful due to long incubation time (one month) of this parasite in the culture medium.
4. **Serological diagnosis:** ELISA and IFA are widely used for diagnosis of acute disease.
5. **Molecular diagnosis:** A PCR technique is highly sensitive for detection of this parasite.

Disease Management in Humans

1. **Control of triatomine bugs:** It is the most important measure for prevention of disease. Use of insect repellents in bug-infested endemic areas is important in this regard.
2. **Improvement in living standards:** Improvement of housing and living conditions for people in rural areas to avoid triatomine bug-infestations is important to prevent the infection in humans.
3. **Control on travel:** Persons should avoid traveling in potentially bug-infested areas.
4. **Treatment:** Patients can be treated with nifurtimox (@ 8-10mg per kg body weight per day, in four divided doses, orally for 3-4 months. An alternative and equally effective drug benznidazole @ 5mg per kg body weight per day, in two divided doses, orally for 1-3 months can also be used.
5. **Surgical intervention:** It is indicated in case of mega-esophagus and mega-colon developed due to this disease.

New World Cutaneous Leishmaniasis

Synonyms

American leishmaniasis, Brazilian leishmaniasis, Naso-oral leishmaniasis, Espundia, Uta, Pain bois, Chiclero ulcer, Forest yaw

Etiology

The disease is caused by *Leishmania mexicana, L. braziliensis complexes, L. peruviana, L. chagasi* and other species.

Epidemiology

The disease occurs in various parts of the world, including South America, Middle East, India, Mediterranean littoral and parts of Africa. The endemic areas of disease extend from southern Texas to northern Argentina. The disease is predominant in people residing in forest and rural areas. Occurrence of disease in temperate climate is associated with returning of persons from endemic areas. Rodents and dogs are reservoirs of these parasites.

Transmission

Leishmaniae are transmitted only by female sandflies (*Lutzomyia* spp.).

Disease in Humans

Cutaneous symptoms are characterized by the development of reddish-blue papules which change into multiple cutaneous nodules and ulcerate. Cutaneous symptoms similar to those of old world cutaneous leishmaniasis but lesions are extensive and slowly resolved.

Diagnosis

1. **Cultural examination:** Organisms can be grown on Novy-McNeal-Nicolle (NNN) medium. This is a common and world known medium for isolation of *Leishmania*. This consists of two phases, blood agar base (part A) and Locke's solution (part B). Blood agar base is highly nutritious medium that supports the growth of fastidious organisms like *Leishmania* and *Trypanosoma*. The specimens are inoculated into the liquid phase of the biphasic medium and incubated. The amastigotes transform to promastigotes in about 24 hours.

2. **Microscopic examination:** Touch preparations and tissue sections from the border of the cutaneous lesions are taken for skin biopsy and examined for parasite. Direct smear examination can also be performed from peripheral blood of the patient.

3. **Serological diagnosis:** Anti-leishmanial antibodies can be detected by IHA and IFT.

4. **Molecular diagnosis:** Confirmatory diagnosis can be made by using PCR technique.

Disease Management in Humans

1. **Control of sandflies:** Ensure protection from bites by the sandflies by using insecticides, repellents and protective clothing.

2. **Treatment:** Patients can be treated using pentavalent antimony compounds such as sodium stibogluconate @ 10mg per kg body weight, twice a day, i.v. or i.m. for 28 days.

Old World Cutaneous Leishmaniasis

Synonyms

Oriental sore/boil, Delhi boil, Baghdad boil, Tropical boil, Lahore ulcer, Aleppo

Old world cutaneous leishmaniasis is caused by protozoan parasites that affect mainly the skin.

Etiology

The disease is predominantly caused by *Leishmania tropica minor* (dry form), *L. tropica major* (wet form) and *L. aethiopica* (chronic oriental sore or old world diffuse cutaneous leishmaniasis).

Epidemiology

The disease is endemic in Eastern Mediterranean, East Africa and Southwest Asia. Increased tourism has resulted the cutaneous leishmaniasis a worldwide disease. Generally one of the both leishmaniasis (cutaneous and visceral) dominates regionally. In rural areas *L. tropica major* is the predominating parasite. The reservoirs of *Leishmania tropica* are rodents.

Transmission

The transmission of *Leishmania tropica minor* and *L. tropica major* occurs mainly by the sandflies (*Phlebotomus sergenti* and *P. papatasii*) bites. The transmission of *Leishmania aethiopica* occurs by *Phlebotomus longipes* and *Phlebotomus pedifer*.

Disease in Humans

The classical cutaneous leishmaniasis (*Leishmania tropica*) is characterized by development of reddish-blue papules which change into multiple cutaneous nodules and later on ulcerate. Lesions cause itching. Leishmaniasis caused by *Leishmania tropica* and *L. aethiopica* induce multiple cutaneous lesions with severe inflammation and exudation causing "moist" ulcers. The lesions occur mainly on the exposed body parts like face, ears, nose and upper and lower extremities. Diffuse leishmaniasis (caused by *L. aethiopica*) is characterized by development of papule without ulceration but satellite lesions develop around it.

Diagnosis

1. **Cultural examination:** Organisms can be grown on NNN agar.
2. **Microscopic examination:** Touch preparations and tissue sections from the border of the cutaneous lesions are taken for skin biopsy and examined for amastigotes.
3. **Serological diagnosis:** Anti-leishmanial antibodies can be detected by ELISA, IFA and direct agglutination.

Disease Management in Humans

1. **Control of vectors:** Sandflies are the vectors which can be controlled by spraying of insecticides like lindane (hexachlorocyclohexane) in

breeding places and human residences. Use of sandfly repellents (hexamethylbenzamide or diethyl toluamide) can prevent bites by them.

2. **Control of reservoirs:** The rodents are the reservoirs. They can be controlled by using traps, rodenticides *etc.* Infected dogs also act as reservoir and should also be destroyed.

3. **Vaccination:** Vaccines can be used but are not fully effective due strains specificities (variants) in different areas of endemicity.

4. **Treatment:** The drugs of choice are pentavalent antimony compounds such as sodium stibogluconate or meglumine antimonite (@ 10mg per kg body weight, twice a day, i.m. or i.v. for 28 days.

Visceral Leishmaniasis

Synonyms

Kala-azar, Black disease

Visceral leishmaniasis is a systemic infection caused by the members of *Leishmania donovani* group.

Etiology

The disease is usually caused by *Leishmania donovani donovani*. It is also caused by *L. donovani chagasi* in Central and South America and some Caribbean islands. *L. donovani infantum* infection is endemic in Mediterranean countries.

Epidemiology

The disease affects humans of all age groups suffering from immunosuppressive disease. The main reservoirs are humans, dogs and rodents. In India and East Africa, the reservoirs of kala-azar are humans (Anthroponotic forms). In Mediterranean-Central Asiatic and South America, the reservoirs of visceral leishmaniasis are canines such as dog, fox and jackals. In Africa the main reservoirs of kala-azar are rodents.

Transmission

The disease is transmitted by sandflies (*Phlebotomus argentipes, P. papatasii,*

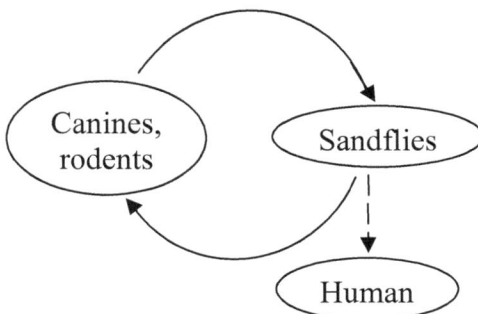

Figure 46.4: Infection Cycle of Visceral Leishmaniasis.

Lutzomyia intermedius and L. longipalpis) bites. Only female sandflies suck the blood. Sandflies bite mainly after sunset and shortly before sunrise. During blood sucking on vertebrates the sandflies ingest free amastigotes or parasitized host cells. High humidity, temperature and availability of organic substrate favour the breeding of sandflies. Their activities are high under high humidity and calm weather conditions.

Disease in Humans

The disease is characterized by development of primary skin lesion (leishmanioma) which occurs after bite by infected sandfly. Initially, itching papule with an erythematous rim develops at the site of bite. Acute form is characterized by sudden onset of fever and chills. Intermittent or remittent fever is observed in most of the cases. Other symptoms are anorexia, weakness, anemia and edema. Chronic kala-azar is characterized by fever, massive hepatomegaly, generalized peripheral lymphadenopathy, pneumonia, diarrhea and extensive growth of eyelashes.

Diagnosis

1. **Clinical examination:** Cases with fever, progressive weight loss, massive hepatomegaly, pancytopenia, hypergammaglobulinemia and hyperalbuminemia in endemic areas are pathognomonic.

2. **Microscopic examination:** *Leishmania* spp. can be detected in peripheral blood by blood smear examination.

3. **Serological diagnosis:** Antibodies can be detected by ELISA. This method has limitations due to cross-reactivity with other organisms like *Mycobacterium leprae, Plasmodium* spp., *Schistosoma* spp., *Trypanosoma cruzi* and *Toxoplasma gondii.*

4. **Molecular diagnosis:** PCR method can be used for confirmatory diagnosis of *Leishmania* spp.

Disease Management in Humans

1. **Control of vectors:** Sandflies are the vectors which can be controlled by spraying of insecticide like lindane (hexachlorocyclohexane) on breeding places and human residences. Use of sandfly repellents (hexamethylbenzamide or diethyl toluamide) and mosquito net to prevent bites by them. Organic wastes like refuge, animal manure and foliage should be regularly removed to prevent breeding of sandflies.

2. **Control of reservoirs:** Infected dogs act as reservoir. Dogs in the endemic areas should be examined and destroyed if found infected.

3. **Treatment:** The drugs of choice are pentavalent antimony compounds such as sodium stibogluconate (@ 10mg per kg body weight, twice a day, i.m. or i.v. for 28 days). In case of relapse the therapy should be continued for 40 days.

Chapter 47

Zoonoses Caused by Trematodes

Trematode parasites are commonly known as flukes. They are flatworms belonging to class Trematoda of the phylum Platyhelminthes. They have short unsegmented flattened bodies, covered by a protective cuticle, and bear hooks or suckers for attachment to their host. Trematodes are classified into two main groups, *viz.*, the Monogenea, which are mostly parasite of the skin and gills of fish, and Digenea. These parasitize the intestine, billiary system, vascular system and urogenital tract of vertebrates, and include several important pathogens of domestic animals and humans. There are several species of trematodes which have zoonotic importance. In most of the zoonotic trematode infections, human acts as definitive host. Transmission of infections occurs mostly through intermediate hosts.

Cercarial Dermatitis

Synonyms

Swimmer's dermatitis, Swimmer's itch, Cercarial itch, Calm digger's itch, Gulf coast itch, Rice filed itch.

Cercarial dermatitis occurs due to infection with several parasites belong to the family Schistosomatidae.

Etiology

The disease is caused by cercariae of several species belong to the genera Schistosomatids such as *Austrobilharzia, Trichobilharzia, Gigantobilharzia* and *Schistosomaticum.* Freshwater snail is the intermediate host, while freshwater fowl as a final host. Mammals also act as final host of some parasites.

Epidemiology

The disease occurs worldwide with the exception of arctic regions.

Transmission

Snails are infected with faeces of infected animals (cattle and buffaloes). Cercariae released from snail and infect animals, waterfowl and accidently humans.

Disease in Humans

The disease is characterized by development of reddish spot of several millimeters on skin at the site of entry of parasite, itching, malaise *etc.*

Diagnosis

The disease can be diagnosed by knowing the history of bathing in the surface water particularly in lakes in wooded areas and working in flooded rice fields and clinical examination. Disease can also be diagnosed by immunofluorescence assay.

Disease Management in Humans

1. **Prevention of contact with contaminated water:** Avoid bathing or swimming in lake and coastal areas infested with parasite.
2. **Control of ducks and other waterfowls:** Keep away the ducks or other waterfowl from lake infested with parasite in order to interrupt the life cycle of the parasite.
3. **Control of snail populations:** Reduce the vegetation at the sore of bathing lake in order to reduce the snail population.
4. **Treatment:** Antipruritic and antiphlogistic powder or lotion can be applied on affected areas of skin.

Clonorchiasis

Clonorchiasis is a liver disease caused by the Chinese liver fluke affecting mainly bile ducts.

Etiology

The disease is caused by a trematode parasite *Clonorchis sinensis.* It is lancet-shaped and transparent parasite and commonly known as Chinese liver fluke.

Epidemiology

The disease is prevalent in Asian countries particularly China, Japan, Korea, Vietnam and Taiwan.

Transmission

Dogs, cats, pigs and other carnivorous animals including humans are reservoirs of this parasite. The disease is transmitted due to ingestion of freshwater fish (second intermediate host) containing metacercariae. Snail is the first intermediate host.

Disease in Humans

The disease is characterized by fever, cholengitis, hepatomegaly, icterus, abdominal pain, diarrhea *etc.* There may be urticaria, dizziness, tremor and convulsion in heavy infections.

Diagnosis

1. **Microscopic examination:** Eggs of the parasite can be detected in faeces and duodenal secretions.

2. **Serological examination:** ELISA can be used for diagnosis of the disease.

Disease Management in Humans

1. **Adequate cooking or frying of fish:** Adequate cooking or frying of fish is important to prevent the spread of infection.

2. **Prevention of contamination of water:** Avoid introduction of human faeces into fish waters.

3. **Control of snail population:** The disease can be prevented by controlling snail population.

4. **Treatment:** Disease can be treated with praziquantel @ 25mg per kg body weight, three times a day, orally, for 2 days. Albendazole @ 10mg per kg body weight, daily, orally, for one week can be used as an alternative drug.

Fascioliasis

Fascioliasis is a liver disease of animals and humans and caused by liver flukes.

Etiology

The disease in humans is mainly caused by *Fasciola hepatica*. *F. gigantica* rarely causes disease in humans.

Epidemiology

The disease occurs worldwide. Moist conditions and high rainfall favour the development of intermediate host (amphibious snail of the genus *Lymnaea*). The final hosts are ruminants, especially sheep.

Transmission

In human beings, the disease is transmitted by ingestion of metacercariae adhering to plants and their fruits.

Disease in Humans

The disease is characterized by fever, anemia, peritonitis, fibrosis and calcification of bile ducts and icterus.

Diagnosis

1. **Microscopic examination:** Microscopic examination of faeces may be helpful in the diagnosis of disease. However, this method may require repeated examination because eggs are not continuously shed in the faeces.

2. **Serological examination:** ELISA method can be used for detection of excretory-secretory *F. hepatica* antigens.

3. **Radiological examination:** CT scan, MRI and X-rays can also be used for diagnosis of disease.

Disease Management in Humans

1. **Precautions and hygienic measures:** Avoid chew on grass blades. Avoid consumption of watercress from natural habitat. Avoid consumption of unboiled or unfiltered surface water.

2. **Treatment:** The disease can be treated with triclabendazole @ 10mg per kg body weight, daily, orally, in combination with bithionol @ 20mg per kg body weight, twice a day, every second day, for 2 weeks.

Fasciolopsiasis

Fasciolopsiasis is an intestinal disease of humans caused by giant intestinal flukes.

Etiology

The disease is caused by *Fasciolopsis buski*. The parasite is commonly known as giant intestinal fluke.

Epidemiology

The disease is prevalent mainly in Southeast Asia particularly China, Taiwan, Vietnam, Indonesia, Thailand, India and Bangladesh. Pigs and humans are the main reservoirs of the parasite. The disease prevalence is high in children.

Transmission

Ingestion of metacercariae adhering to water plant and its fruits leads to infection in the humans. Snail is the intermediate host.

Disease in Humans

The disease is characterized by inflammation and ulcers in the intestine, diarrhea, gastric pain, edema, ascites, weight loss, anemia *etc.*

Diagnosis

Parasite's eggs (operculated eggs filled with yolk material) can be demonstrated in the fecal samples.

Disease Management in Humans

1. Avoid consumption of raw water nuts and chestnuts.
2. Prevention of contamination of water with pig faeces.
3. Control of snail population.
4. Treatment with praziquantel (@ 25mg per kg body weight, three dosages in one day).

Paragonimiasis

Synonym

Pulmonary distomatosis

Etiology

The disease is caused by *Paragonimus westermani*. The parasite is commonly known as lung fluke.

Epidemiology

The disease is predominant in Southeast and Central Asia and South America.

Transmission

The disease is transmitted by ingestion of meat of freshwater crabs and other crustaceans (second intermediate host) containing cercariae. Snail is the first intermediate host.

Disease in Humans

The disease is characterized by pleurisy, pneumonia, chest pain, fever, peritonitis *etc.*

Diagnosis

1. **Microscopic examination:** The disease can be diagnosed by demonstration of eggs of the parasite in the sputum and faeces.
2. **Serological examination:** When sputum is negative for parasite's eggs, the ELISA can be helpful for diagnosis of the disease.

Disease Management in Humans

1. Avoid consumption of raw or undercooked crustaceans.
2. Control of snail population.
3. Treatment with praziquantel @ 25mg per kg body weight, three times a day, orally, for 2 day.

Schistosomiasis

Synonym

Bilharziosis

Etiology

Intestinal schistosomiasis is caused by *Schistosoma mansoni and S. japonicum*, while schistosomiasis of urinary bladder is caused *S. haematobium*.

Epidemiology

S. mansoni and *S. haematobium* are prevalent in Africa, while *S. japonicum* in Asia.

Transmission

The disease is transmitted by contact of skin with cercariae contaminated water. Snail is the intermediate host.

Disease in Humans

Intestinal schistosomiasis is characterized by skin lesions at the site of cercariae penetration, fever, headache, constipation, bloody diarrhea, lymphadenopathy, splenomegaly, hepatomegaly, fibrosis of intestinal wall and liver and liver cirrhosis. Schistosomiasis of urinary bladder is characterized by pruritus at the site of cercariae penetration, fever, headache, generalized pain, vomiting, hematuria, fibrosis and calcification of the wall of the urinary bladder and obstructive uropathy.

Diagnosis

1. **Microscopic examination:** Schistosomal eggs can be demonstrated in the urine or faeces.

2. **Serological examination:** The disease can also be diagnosed by ELISA, IHA and IFA.

3. **Radiological examination:** Ultrasonography is useful for detection of Schistosoma-induced organ lesions.

Disease Management in Humans

1. Avoid contact of surface water in endemic zones.

2. Prevention of contamination of water with human faeces and urine.

3. Control of snail population.

4. Treatment with praziquantel.

Chapter 48

Zoonoses Caused by Cestodes

Cestodes are also known as tapeworms. They are flatworms belong to the class Cestoda and the family Platyhelminthes. The adult stage of cestodes parasites are found usually in intestine and their larval stages in the body tissues of the hosts. Several genera are of zoonotic significance, including *Dipylidium* in dogs, *Taenia* in cattle, pigs and humans. Most of the infections require intermediate hosts. Adult parasites cause little harm to the hosts. The larvae are usually more damaging than the adults, causing diseases such as coenurosis, cysticercosis, sparganosis and hydatid disease in their intermediate hosts.

Echinococcosis

Synonyms

Hydatidosis, Hydatid disease

Echinococcosis is a parasitic disease that occurs in two main forms in humans that is cystic echinococcosis (also known as hydatidosis) and alveolar echinococcosis, caused by the tapeworms *Echinococcus granulosus* and *E. multilocularis*, respectively. Dogs, foxes and other carnivores harbour the adult parasite in their intestine and pass the parasite eggs in their faeces. If the eggs are ingested by humans, they develop into larvae in several organs, more frequently the liver and lungs. Both cystic and alveolar echinococcosis are characterized by asymptomatic incubation periods that can last many years until the parasites grow to an extent that triggers clinical signs.

Etiology

Hydatid disease is caused by *Echinococcus* spp., a small tapeworm belongs to the family Taenidae. Echinococcosis occurs in the following forms-

1. **Cystic echinococcosis (hydatidosis or hydatid disease):** This form is caused by infection with *E. granulosus*. It is found in the small intestine of

carnivores and the metacestode (hydatid cyst) in ungulates and humans. In humans, the infective larvae cause "unilocular" type of echinococcosis.

2. **Alveolar echinococcosis:** This form is caused by infection with *E. multilocularis*. It is perpetuated mainly by a sylvatic cycle involving foxes and cricetid rodents. A major source of infection in humans is through fruits and vegetables that have been contaminated by the faeces of foxes and occasionally cats and dogs. Rodents are intermediate host. In humans, the metacestode causes "alveolar or multilocular cysts".

3. **Polycystic echinococcosis:** This form is caused by infection with *E. vogeli*. It is found in bush dogs and domestic dogs which are definitive hosts. Pacas (*Cuniculus paca*) and possibly other rodents act as intermediate host.

4. **Unicystic echinococcosis:** This form is caused by infection with *E. oligarthus*. It is found in puma and cat which are definitive host. Rodents are intermediate host.

Epidemiology

E. granulosus occurs worldwide with highly endemic areas in the former USSR, the Mediterranean countries and area of Africa, Latin America and Australia. *E. multilocularis* is found in the Northern hemisphere, Central and Eastern Europe and particularly in Canada and former USSR. *E. vogeli* occurs in Central and Northern South America. *E. oligarthus* occurs in Central and South America. More than 1 million people are affected with echinococcosis at any one time.

Transmission

1. **Ingestion:** Human infection usually occurs due to ingestion of food, unwashed vegetables or water contaminated with infected dog faeces. Infection may also occur due to handling or playing with infected dogs (hand to mouth transfer of eggs). Carnivores get infection by consuming viscera containing hydatid cysts.

2. **Inhalation:** Infection may be transmitted due to inhalation of dust contaminated with infected eggs.

Note: Human is accidental dead end host in *Echinococcus* infection. This infection is not transmitted from person to person.

Life Cycle

The adult tapeworm lives in the small intestine of dogs (definitive host) for 2-4 years. The eggs are voided in the faeces and contaminate the soil, water and pasture. Cattle, sheep and some other herbivores (intermediate host) become infected by grazing contaminated pastures. Ingested eggs hatch in the intestine and onchosphere penetrates the intestinal mucosa and migrate to various organs via circulation. Most frequently they lodge in the liver, lungs and brain and develop into hydatid cysts. The life cycle is completed when cattle and sheep viscera containing hydatid cysts are consumed by dogs. The infected dogs begin to pass eggs of the parasite approximately 7 weeks after infection. Human does not harbour the adult worms.

Note: In *Echinococcus* infection, the gravid proglottids usually disintegrated in the intestine, so that only eggs and not proglottids are found in the faeces of dogs. While in taeniasis, the gravid proglottids are released in the faeces.

Disease in Animals

Pathogenicity of hydatid disease depends on the severity of infection and the organ involved. Clinical manifestations are not usually seen despite heavy infection. Except in cases where cysts become too large and exert pressure on visceral organs, the disease remains subclinical.

Disease in Humans

The two most important forms of the disease in humans are cystic echinococcosis (hydatidosis) and alveolar echinococcosis. The clinical manifestations are seen after several years of exposure to infection. The cysts grow slowly from 5-20 years before they are diagnosed. The size of cyst may vary from pin head to that of small football. More than 70 per cent of the cysts are located in the right lobe of liver and the rest in lungs, brain, peritoneum and long bones. Clinical symptoms do not occur if the size of cysts is small. Large cysts cause pressure symptoms (jaundice in liver cysts). In vital organs, they may cause severe symptoms and death.

Diagnosis

Under normal conditions of fecal examination the eggs of *Echinococcus* cannot be differentiated from those of *Taenia* spp. The disease can be diagnosed by the following methods:

1. **Radiological examination:** Ultrasonography, X- rays, CAT scan and MRI can be used for diagnosis of disease. Ultrasonography is the imaging technique of choice for the diagnosis of both cystic echinococcosis and alveolar echinococcosis. This technique is usually complemented or validated by computed tomography (CT) and/or MRI scans.

2. **Serological examination:** ELISA has been useful for diagnosis of *Echinococcus* infection.

3. **Intradermal test (Casoni test):** This test is still in wide use.

Disease Management in Animals

1. **Hygienic measures:** High standards of hygiene should be adopted at animal farms and slaughterhouses. Dogs should not be allowed to access the infected viscera (hydatid cysts) of cattle, sheep *etc.*

2. **Chemoprophylaxis:** Chemoprophylaxis with anthelmintics on mass basis in dogs is useful for prevention of disease.

3. **Treatment:** Praziquantel @ 5 mg per kg body weight in dog can remove the parasites from intestine.

Disease Management in Humans

1. **Control of dogs:** Stray dog elimination is important in controlling hydatid disease. There should be periodic examination of dog faeces and administration of anthelmintic such as aricoline hydrochloride. Dogs should also be prevented from access of raw offal at slaughterhouses and on farms and dead animals.

2. **Surveillance:** Surveillance for cystic echinococcosis in animals is difficult because the infection is asymptomatic in livestock and dogs. Periodic deworming of dogs, improved hygiene in the slaughtering of livestock (including the proper destruction of infected offal), and public education campaigns have been found to lower the occurrence of disease in humans. Alveolar echinococcosis prevention and control is more complex as the cycle involves wild animal species as both definitive and intermediate hosts. Regular deworming of domestic carnivores that have access to wild rodents is helpful in reducing the risk of infection in humans. Deworming of wild animals with anthelmintic baits has been resulted in significant reductions in prevalence of alveolar echinococcosis. Culling of foxes and unowned free-roaming dogs is applicable but not efficient.

3. **Health education:** Health education particularly butchers, dog owners, animal breeders and shepherds may be effective in prevention of disease.

4. **Vaccination:** Vaccination of sheep with an *E. granulosus* recombinant antigen (EG95) is important in prevention and control of the disease.

5. **Treatment:** Both cystic echinococcosis and alveolar echinococcosis are often expensive and complicated to treat, sometimes requiring extensive surgery and/or prolonged drug therapy. Surgical removal of cyst may be risky if even single cyst accidentally penetrates may lead to shock and death. However, four options exist for the treatment of cystic echinococcosis.

    i. Percutaneous treatment of the hydatid cysts with the PAIR (Puncture, Aspiration, Injection, Re-aspiration) technique.

    ii. Surgery.

    iii. Anti-infective drug treatment.

    iv. Watch and wait.

The method of treatment is based on the ultrasound images of the cyst, following a stage-specific approach, and also on availability of the medical infrastructure. For alveolar echinococcosis, early diagnosis and radical (tumour-like) surgery followed by anti-infective prophylaxis with albendazole remain the key elements. If the lesion is confined, radical surgery offers cure.

Taeniasis

Taeniasis is a group of cestode infections of zoonotic importance. Taeniasis is acquired by humans through the ingestion of tapeworm larval cysts (cysticerci) in undercooked pork or beef. Human tapeworm carriers excrete tapeworm eggs in their faeces and contaminate the environment when they defecate in the open. The disease is characterized by nausea, abdominal pain and weakness. The larvae of *Taenia* (*Cysticercus*, cyst or bladder worm) affect many animal species and human and produce disease called cysticercosis.

Etiology

The disease is caused by *Taenia* spp. *T. saginata* occurs in the small intestine of humans (the only definitive host) and metacestode (*Cysticercus*) is found in cattle. *T. solium* is found in pigs and human beings act as definitive host.

Epidemiology

Taenia saginata has a cosmopolitan distribution and infection is particularly important in Africa, South America and in some Mediterranean countries. *Taenia solium* is endemic in Latin America, South Africa and Southeast Asia. *T. solium* is the cause of 30 per cent of epilepsy cases in many endemic areas where people and roaming pigs live in close proximity. The infection is common in areas where insanitary conditions prevail and pigs run loose scavenging for food, with ready access to human faeces.

Transmission

The contaminated feed and water are the main sources of infection in animals, while meat infected with cysticerci, contaminated food, water and vegetables are the main sources of infection in human beings. There are following modes of transmission of taeniasis.

1. **Ingestion:** This is the common mode of transmission of taeniasis. The disease may be transmitted due to ingestion of infected meat and contaminated food and water.

2. **Re-infection:** This mode of transmission is rare but may occur due to transport of eggs from the bowel to the stomach by retro-peristalsis.

Life Cycle of *T. saginata*

The proglottids containing eggs are released along with the human faeces which contaminate pastures and may be grazed by cattle. Following ingestion the onchosphere hatched out and activated under the influence by gastric and intestinal juice and penetrate the intestinal mucosa and reach to the circulation. The onchosphere develops into embryos and are disseminated throughout the body and develop in skeletal and cardiac muscles and organs. *Cysticercus bovis* develops and becomes infective in about 10 weeks and remain viable for up to 9 months or longer. Human is infected by ingestion of infected (measly) beef. Gravid proglottids begin to be passed by man approximately 100 days after infection. The life cycle of *Taenia solium* is similar to that of *Taenia saginata* except that pig acts as intermediate host. Cysticerci develop primarily in cardiac and skeletal muscles. The fully developed *Cysticercus* is infective after about 9-10 weeks. The life cycle is completed and humans are infected when they eat raw or undercooked infected (measly) pork.

Disease in Animals

The clinical manifestations are not normally seen in infected animals. In case of heavy infection, cattle may show fever, anorexia, pericarditis and embolism. The parasite is of great importance as a cause of financial loss to the meat industries.

Disease in Humans

Clinical manifestations are abdominal discomfort, anorexia and chronic indigestion. Sometimes straying of proglottids may cause appendicitis or cholengitis. The most serious complication of *T. solium* infection is cysticercosis. Cysticercosis involving central nervous system is known as neurocysticercosis which is a serious threat to the individual. There may be development of mechanical pressure, obstruction or inflammation in neurocysticercosis. This may lead to headache, blindness, meningitis, epilepsy, intracranial hypertensive syndrome, hydrocephalus, dementia and death.

Diagnosis

1. **Serological examination:** ELISA and passive haemagglutination are useful for detection of infection.
2. **Radiological examination:** This is helpful in supporting in clinical and laboratory diagnosis.

Disease Management in Animals

1. **Hygienic measures:** Improved sanitation and improved pig husbandry is helpful in preventing the spread of disease.
2. **Meat inspection:** Regular meat inspection at slaughterhouses must be carried out to interrupt the transmission chain of disease.
3. **Treatment:** Praziquantel, albendazole or mebendazole can be used for treatment of the disease.

Disease Management in Humans

1. **Hygienic measures:** Personal and environmental hygiene should be adopted properly to prevent the spread of infection.
2. **Meat inspection:** Improved meat inspection and processing of meat products must be ensured in order to prevent the transmission of the disease.
3. **Heat treatment of meat:** Pork or beef should be properly cooked before their consumption.
4. **Health education:** Health education to the public about source of infection, method of transmission and severity of disease may be of some measures to prevent the spread of disease.
5. **Treatment:** Taeniasis can be treated with praziquantel (@5-10 mg per kg body weight, single-administration) or niclosamide (adults and children over 6 years: 2 g, single-administration after a light breakfast, followed

after 2 hours by a laxative; children aged 2-6 years: 1 g; children under 2 years: 500 mg). Currently there are no standard treatment guidelines for neurocysticercosis and treatment has to be tailored to the individual case. Since the destruction of cysts may lead to an inflammatory response, treatment of active disease may include long courses with praziquantel and/or albendazole, as well as supporting therapy with corticosteroids and/or anti-epileptic drugs, and possibly surgery. The dosage and the duration of treatment can vary greatly and depend mainly on the number, size, location and developmental stage of the cysts, their surrounding inflammatory edema, acuteness and severity of clinical symptoms.

Chapter 49

Zoonoses Caused by Nematodes

Nematodes are also known as roundworms. They are invertebrate animals belonging to the phylum Nematoda. They typically have a smooth cylindrical unsegmented body tapered at both ends and protected by an outer cuticle. The life cycle may be direct (*e.g., Ancylostoma* spp. and *Ascaris* spp.) or indirect (*e.g., Habronema* spp.) involving intermediate hosts. Alternatively, the eggs may be taken up by a transport host (*e.g.,* earthworm, as in *Syngamus* spp.). They are found virtually in all habitats and include both free-living and parasitic forms. Parasitic forms cause diseases in animals and humans. The adult parasites occur in various sites in the body. The adult forms of nematodes can reside in the gastrointestinal tract, blood, lymphatic system or subcutaneous tissues. Alternatively, the immature (larval) states can cause disease through their infection of various body tissues.

Anisakiasis

Synonym

Herring worm disease

Herring worm disease is fish zoonosis, caused by larvae of several nematode species and characterized by eosinophilic granuloma or ulcers in intestinal tract of humans.

Etiology

The disease is caused by larvae of anisakids, belong to a nematode family. The larvae of the parasites invade the stomach or intestinal wall and cause granulomas or ulcers. *Anisakis simplex* and *Pseudoterranova* spp. are the most important species.

Epidemiology

The disease has been reported from Asia, America and Europe. Infections with *Pseudoterranova* spp. are limited to Northern Pacific areas.

Transmission

The disease is transmitted due to ingestion of saltwater fish and squids. Crustaceans are the intermediate host. Whales, dolphins and seals are the final host.

Disease in Humans

The incubation period varies from few hours to several days. The disease is characterized by epigastric pain, nausea, vomiting, eosinophilic granuloma and ulcers in the intestinal tract, blood in the faeces and intestinal stenosis. Intestinal perforation may occur. In sensitized hosts, immediate-type hypersensitivity reaction may occur after few hours of ingestion of fish which is manifested by pruritus, acute urticaria or generalized anaphylaxis.

Diagnosis

1. **Biopsy specimen examination:** It is helpful in the diagnosis of disease.
2. **Endoscopy:** The disease can be diagnosed by endoscopic examination of the larvae.
3. **Imaging:** Imaging techniques are also helpful in the diagnosis of disease.
4. **Serology:** ELISA and immunoblotting are important techniques for diagnosis of the disease.

Disease Management in Humans

1. Adoption of hygienic measures in fish industries.
2. Avoid consumption of raw saltwater fish of endemic areas.
3. Extraction of larvae under endoscopic control.
4. Treatment with mebendazole @ 200 mg twice a day, orally for three days.

Filariasis

Zoonotic filariasis is caused by nematode parasites such as *Brugia malayi, B. timori* and *Dirofilaria immitis*. Other filarial parasites such as *Dipetalonema perstans, Dipetalonema streptocerca, Mansonella ozzardi, Onchocerca volvulus* and *Wuchereria bancrofti* also cause disease in humans for which primates may act as reservoir. Infections with *Meningonema peruzzii* are found in Central Africa. *Onchocerca gutturosa* has been reported from Japan.

Brugia Filariasis

Synonym

Lymphatic filariasis

Brugia filariasis is generally asymptomatic but may be associated with obstructive changes in the lymphatic system and cause elephantiasis.

Etiology

Brugia filariasis is caused by *Brugia malayi* and *B. timori*, which are of zoonotic importance.

Epidemiology

The disease caused by *Brugia malayi* is endemic in Asia and has been reported from India, Burma, Thailand, Vietnam and Philippines. There are two strains of *B. malayi*. The nocturnal periodic strain is found in peripheral blood only at night, while during the day they accumulate in the lung capillaries. The sub-periodic strain is found in the peripheral blood. This strain is found in the rainforest areas of Southeast Asia. *Brugia timori* is found in Indonesia. The important reservoirs are dog, cats and wild felids.

Transmission

The disease is transmitted by infected mosquito bite. The important vectors are *Aedes* spp. and *Mansonia* spp. Mosquitoes transmit the third stage larvae of the parasite. The adult stage of the parasite lives in the lymph vessels and lymph nodes.

Disease in Humans

The disease may be asymptomatic in most of the cases. The disease is characterized by lymphangitis and lymphadenitis mostly in the legs and groin, fever, headache, backache and elephantiasis in the later stage of disease which is manifested by edema in the lower part of the legs.

Diagnosis

1. **Blood smear examination:** Microfilariae can be demonstrated in the blood sample using Giemsa stain.
2. **Serological examination:** High titre of IgG4 antibodies is indicative of infection. Serological diagnosis can also be performed by demonstration of circulating filarial antigens.

Disease Management in Humans

1. Control of mosquitoes.
2. Treatment with diethylcarbamazine @ 50 mg on day one, 50 mg thrice on day two, 100 mg thrice on day three and 2mg per kg body weight thrice a day from days 4 to 21, orally.
3. Surgical removal of nodules and parasites.

Dirofilariasis

Dirofilariasis occurs due to infection with *Dirofilaria* spp. The disease affects mainly the lungs and subcutaneous tissues, and occasionally conjunctiva.

Etiology

The disease is caused by *Dirofilaria immitis* (heartworm of dogs). The disease caused by *D. repens, D. tenuis, D. ursi* and *D. striata*, develop in dogs, raccoons, bears and felids, respectively, which are important reservoirs for maintaining the infection.

Epidemiology

D. immitis occurs worldwide. The disease is prevalent in United States, Central and South Americas, Spain, Italy and East Asia.

Transmission

The disease is transmitted by infected mosquitoes and simulia (blackflies) during blood sucking. The adult parasite is found in the heart and pulmonary arteries of their final hosts. The final host is canine.

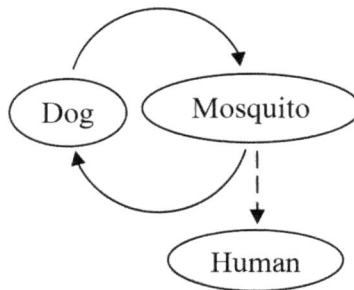

Figure 49.1: Infection Cycle of Dirofilariasis.

Disease in Humans

The disease is asymptomatic in most of the human cases. *D. immitis* usually inhabits the parenchyma of the lung and the disease is clinically characterized by chest pain, coughing, hemoptysis, localized vasculitis, pulmonary infarcts and granuloma around the worms. Other species of *Dirofileria* settle in the subcutaneous tissues, occasionally in conjunctiva and cause development of nodules which are painful and itching.

Diagnosis

1. **Radiological examination:** Nodules in the subcutaneous tissues can be examined by X-rays.
2. **Serological examination:** It can be applied if other filarial infections can be excluded.
3. **Excision and demonstration of parasites:** These steps can be used for confirmatory diagnosis of dirofilariasis.

Disease Management in Humans

1. **Use of insect repellents:** Insect repellents can be used for protection from biting insects.
2. **Treatment:** The disease can be treated by surgical excision of parasites.

Larva Migrans Cutanea

Synonyms

Cutaneous larva migrans, Creeping eruption

Larva migrans cutanea is mostly acute syndrome of the skin caused by migrating larvae of nematode parasites. Humans are accidental hosts.

Etiology

The disease is caused by invasion of skin by larvae of nematode parasites such as *Ancylostoma braziliense, A. caninum, A. tubaeforme, Bunostomum phlebotomum* and *Uncineria stenocephala*.

Epidemiology

The hookworms (Ancylostomatidae) and *Strongyloides* spp. occur in animals worldwide. *A. braziliense* is found predominantly in tropical and subtropical areas. The incidence of human infection is high on urban beaches in South America and South Africa due to pollution by infected dog faeces.

Transmission

Humans are infected percutaneously due to penetration of bare skin by larvae of the parasites. The moist and warm environmental conditions are conducive for the development of larvae.

Disease in Humans

There is development of papule at the site of penetration of the larval parasites. This is followed by migration of larvae between the corium and stratum granulosum of the skin. The burrows appear on the skin surface as elevated alterations with erythema, edema and crusts. This condition causes itching and pricking pain.

Diagnosis

The disease can be diagnosed on the basis of clinical manifestation (migrating track of the larvae in the skin).

Disease Management

1. **Hygienic measures:** Dogs and cats should be kept away from beaches and playgrounds.
2. **Treatment of dogs and cats:** Dogs and cats should be examined for parasites and treated properly.
3. **Treatment in humans:** Ivermectin is more effective drug and can be used @ 0.2 mg per kg body weight in a single dose, orally. Albendazole can be used as an alternative drug @ 200 mg, twice a day, orally, for 5 days. Antipruritic drugs can be used to reduce itching.

Larva Migrans Visceralis

Synonym

Visceral larva migrans

Larva migrans visceralis is a syndrome caused by the invasion of inner organs by nematode larvae. Human acts as paratenic host.

Etiology

The disease is caused by larvae of ascarid worm of dog (*Toxocara canis*) and ascarid worm of cat (*T. mystax*). The disease is also caused by ascarid worm of raccoon (*Baylisascaris procyonis*).

Epidemiology

The disease caused by *T. canis* in carnivores occurs worldwide. Human infection also occurs worldwide. Antibodies against *T. canis* are found worldwide in 2-14 per cent of the human population. The prevalence of the disease is higher in children.

Transmission

The disease is transmitted by ingestion of food and water contaminated with embryonated eggs (larvae-containing eggs) from infected dog faeces. After ingestion, the larvae hatch in the intestinal wall and spread throughout body via blood circulation.

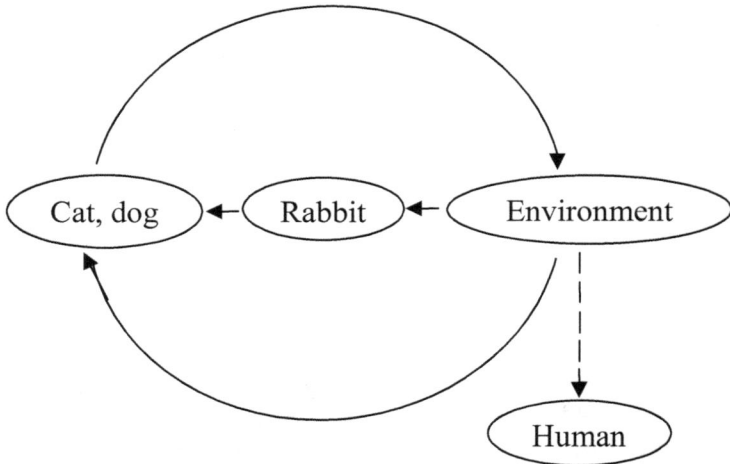

Figure 49.2: Infection Cycle of Larva Migrans Visceralis.

Disease in Humans

The clinical manifestations are seen in two forms that is visceral larva migrans (VLM) and ocular larva migrans (OLM). The VLM is found mainly in the children and characterized by fever, abdominal pain, cough, eosinophilia and leukocytosis.

The persistent re-infections are characterized by hepatomegaly, lymphadenopathy, pneumonia and urticaria. The invasion of CNS is characterized by seizures. OLM occurs due to migration of single larva in one eye and characterized by diffuse unilateral subacute neuroretinitis syndrome.

Diagnosis

1. **Clinical examination:** Clinical signs are eosinophilia, leukocytosis, hepatomegaly and swelling of abdominal lymph nodes.
2. **Serological diagnosis:** ELISA is the choice of serological method of the disease diagnosis.
3. **Ophthalmoscopy:** It is required to visualize the presence of larva in the eye.

Disease Management

1. **Hygienic measures:** Children should be kept away from contaminated playgrounds.
2. **Treatment of dogs and cats:** They should be examined for parasites and treated properly with anthelmintics.
3. **Treatment in humans:** Albendazole can be used @ 400 mg, twice a day, orally, for 5 days. Anthelmintics and corticosteroids are required for the treatment of OLM.

Trichinellosis

Synonyms

Trichinosis, Trichiniasis

Trichinellosis is a nematode infection affecting intestine and striated muscles in the host. *Trichinella* spp. has been found throughout the world which causes human trichinellosis. It causes public health hazard by affecting human patients and also represents an economic problem in swine production and food safety.

Etiology

The disease is caused by *Trichinella spiralis*, *T. pseudospiralis* and other *Trichinella* spp.

Epidemiology

Trichinella spp. occurs worldwide. *T. nativa* is found in the arctic and subarctic zones, while *T. britovi* is found in the temperate zone of Palaearctic region. *T. murrelli, T. nelsoni* and *T. papuae* are found in the temperate zone of North America, tropical Africa and New Guinea, respectively.

Transmission

The disease in humans is transmitted due to consumption of infected raw or undercooked meat. Pig is the main host of this parasite. The disease is also transmitted due to consumption of meat from infected horse, camel, dog, sheep, goat, bear and marine animals like walruses.

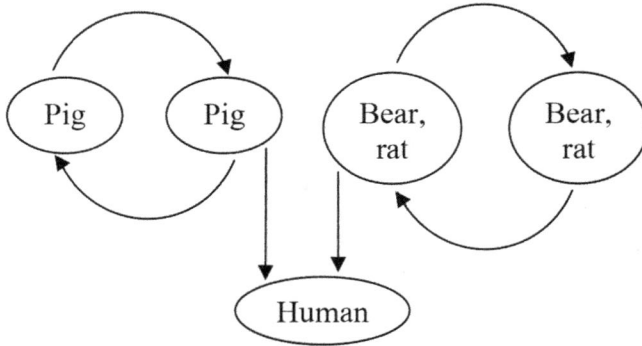

Figure 49.3: Infection Cycle of Trichinellosis.

Disease in Humans

The incubation period may vary from 6 to 40 days. The disease severity depends on the number of larvae ingested and species of *Trichinella*. The disease begins with the intestinal phase which is characterized by nausea, vomiting, epigastric pain and constipation or diarrhea. Later on the visceral phase is characterized by fever, cough, myalgia, conjunctivitis and hemorrhages. Neurological complications are associated with deafness, encephalitis and convulsion.

Diagnosis

1. **Demonstration of larvae in muscles:** *Trichinella* larvae can be directly demonstrated in muscle biopsy specimens.

2. **Serological examination:** Indirect immunofluorescence test employing encysted larvae as antigen and ELISA using extracts or excretory-secretory products of muscle larvae.

Disease Management in Humans

Due to the predominantly zoonotic importance of infection, the main efforts in many countries have focused on the control or elimination of *Trichinella* from the food chain. The disease can be managed in the following ways-

1. Prevention of carnivores from accessing the raw meat in endemic areas.

2. Adequate cooking or freezing (-25°C or below) of meat.

3. Treatment with albendazole (@ 400 mg, twice a day, orally, for 14 days).

Chapter 50

Zoonoses Caused by Arthropods

Arthropods are invertebrate animals possessing segmented body and a hard exoskeleton. The phylum arthropoda includes several classes, *viz.*, Insecta, Archnida, Diplopoda, and Crustacea. Many are of great zoonotic importance. Arthropods like flies, mosquitoes and midges may cause public health problems such as erythema, swelling, urticaria, pruritus and pain. The members of the order Diptra are of zoonotic importance. They have a single pair of functional membranous forewings. The hindwings are reduced to so-called halters. Some important diseases like myiasis and tungiasis are discussed as follows:

Myiasis

Myiasis is the infestation of tissues of humans or vertebrate animals with larvae of the insect order Diptra (flies), which feed on living or necrotic tissues. Infestation starts with first-stage larvae which develop into second and third stage. Larvae can also develop into organic matter.

Etiology

In humans, the myiasis is caused by various genera and species of the order Diptra. Dermal myiasis involves the invasion of the skin which is caused by *Cordylobia anthropophaga* (Tumbu or mange fly), *Cochiomyia hominivorax* (New world screw worm), *Chrysomya bezziana* (Old world screw worm) and *Dermatobia hominis* (human botfly or tropical warble fly). Wound myiasis is caused by *Lucilia sericata, L. cuprina, Caliphora* spp. (blowflies) and *Sarcophaga* spp. (flesh flies). In wound myiasis, both healthy and necrotic tissues can be fed on by the larvae, depending on the conditions and species of fly involved. Creeping myiasis is caused by *Sarcophaga* spp. and *Gasterophilus* spp. Creeping myiasis is a type of cutaneous myiasis involving the migration of fly larvae underneath the skin. Ocular myiasis is caused by *Oestrus ovis* (botflies).

Epidemiology

Myiasis occurs mainly in tropical and subtropical climates. The main contributing factors are probably the higher levels of exposure to myiasis-causing flies due to poorer clothing and hygiene conditions, combined with the increased aggressiveness of myiasis-causing flies in the tropics. However, cases have also been reported from temperate zones. Dermal myiasis is prevalent in Africa, America and Asia, while other myiasis occurs worldwide.

Transmission

Generally female flies of Diptra deposit eggs at the margin of dermal lesions which leads to the infestation in the hosts.

Disease in Humans

The larvae can penetrate several centimeters deep in case of dermal and sub-dermal myiasis. *Dermatobia hominis* causes painful lesions in the skin of head, arm, back, thigh, axila and abdomen. Larvae penetrate the subcutaneous tissues and cause swelling and circuitous tunnel in the lower epidermis in case of creeping myiasis. Creeping myiasis caused by *Gasterophilus* spp. is superficial and present clear, linear serpentine dark tunnels (creeping eruption). *Hypoderma* spp. produces a deeper (subcutaneous), more painful creeping myiasis with discontinuous and transient skin swellings.

Diagnosis

Myiasis can be diagnosed on the basis of clinical examination and demonstration of larvae. A definitive diagnosis is achieved after extraction and identification of the fly larvae. The correct diagnosis of cutaneous myiasis requires knowledge of the patient's exposure history like recent travel to tropical endemic areas, and a degree of clinical suspicion. The furuncular myiasis is characterized by one or more non-healing boil-like lesions on exposed skin with erythema of a few mm to over 2 cm and serosanguineous or seropurulent drainage from a central pore. The local symptoms are pain, tenderness, pruritus, movement inside the lesion or a small, white, worm-like organism protruding from the lesion when pressed laterally. Sometimes, a sharp sting at the site of lesion is felt a few days or weeks before the apparition of symptoms, and is due to the penetration of the larva into the skin.

Disease Management in Humans

1. **Hygienic measures:** Personal and environmental hygiene are important in this regard. Fly breeding habitats can be reduced by managing food residues and garbage containers properly.

2. **Control of flies:** Flies exposure can be minimized by using proper clothing and by maintaining appropriate screening of doors and windows. Sterile male technique is also helpful in controlling flies population.

3. **Removal of larvae:** Larvae can be removed with the help of forceps. Larvae of *Cordylobia anthropophaga* can be expelled from their boils by pressure. Petroleum jelly, paraffin oil, or beeswax can be applied to the opening of

the lesion to asphyxiate the larvae and force it out. Sometimes surgical removal of larvae may be required.

4. **Treatment:** Treatment with antibiotics is required to prevent secondary infection. Intestinal myiasis can be treated with laxatives.

Tungiasis

Tungiasis is a dermatological disease caused by invasion of female sandfleas and characterized by painful lesions on the skin.

Etiology

The disease is caused by female sandflea *Tunga penetrans*. The flea is also referred to as the jigger, chica, nigua, pico, pique or suthi. It burrows into the epidermis and grows in the skin.

Epidemiology

The sandfleas are distributed in tropical and subtropical regions of the world, including Mexico to South America, the West Indies and Africa. The fleas normally occur in sandy climates, including beaches, stables and farms. The disease occurs sporadically in Asia. Sandfleas infect animals, birds and humans.

Transmission

The disease is transmitted due to invasion of skin by adult female sandflea, where it lays eggs for several weeks. Later on adult female sandflea dies. Eggs hatch out and larvae develop in three stages and pupate after two weeks.

Disease in Humans

The disease is characterized by pain at the site of penetration after two days. The skin of any part of the body may be affected. There is severe pruritus, inflammation and lymphangitis when the jiggers grow and mature. Scratching of lesions may cause secondary infections including tetanus and gangrene.

Diagnosis

The disease can be diagnosed by clinical symptoms.

Disease Management in Humans

1. **Hygienic measures:** It must be followed in and around the houses.
2. **Protection from fleas:** Shoes can be used for protection from fleas.
3. **Removal of fleas:** Fleas can be removed by surgical procedure.
4. **Use of insecticides:** Insecticides can be used against adult and larval stages of parasites.
5. **Treatment:** Topical antibiotics can be applied on the lesions to prevent secondary infection.

References

Acha, P. N. and Szyfres, B. 2001. *Zoonoses and Communicable diseases Common to Man and Animals*. 3rd ed. vol. I. Bacterioses and Mycoses. Pan American Health Organization, Washington, DC

Collee, G. J., Fraser, A. G., Marmion, B. P. and Simmons, A. 1996. *Mackie and McCartney Practical Medical Microbiology*. 14th ed. Churchill Livingstone Edinburgh London New York Philadelphia Sydney Toronto

Greenwood. D., Slack, R. C. B. and Peutherer, J. F. 1997. *Medical Microbiology. A Guide to Microbial Infections: Pathogenesis, Immunity, Laboratory Diagnosis and Control.* 15th ed. Churchill Livingstone, New York Edinburgh London Madrid Melbourne San Francisco and Tokyo

Heymann, D. L. 2004. *Control of communicable diseases manual: an official report of the American Public Health Association*. 18th ed. Washington DC, World Health Organization/American Public Health Association.

Krauss, H., Weber, A., Appel, M., Enders, B., Isenberg, H. D., Schiefer, H. G., Slenczka, W., Graevenitz, A. V. and Zahner, H. 2003. Zoonoses: Infectious Diseases Transmissible from Animals to Humans. 3rd ed. ASM Press, Washington, DC.

Park, K. 2002. Park's Textbook of Preventive and Social Medicine. 17th ed. M/s Banarsidas Bhanot Publishers 1167, Prem Nagar, Jabalpur, 482 001 India

Thapaliyal, D. C. 1999. Diseases of animals transmissible to man. 1st ed. International Book Distributing Company (Publishing Division) Chaman Studio Building, 2nd Floor, Charbagh, Lucknow 226 004, U. P., India.

WHO. 2002. WHO Study Group on Future Trends in Veterinary Public Health (1999: Toronto Italy) Future Trends in Veterinary Public Health: report of a WHO Study Group (WHO technical report series: 907.

WHO. 2014. World Health Organization, Regional Office for South-East Asia. A brief guide to emerging infectious diseases and zoonoses. New Delhi: WHO-SEARO.

Appendices

Appendix I

Glossary of Terms

Accidental (incidental) host: A host that harbours an organism that usually does not infect it.

Acute: Relating to a disease or condition with a rapid onset and a short, severe course.

Aerosol: A suspension of fine solid or liquid particles in a gas, usually air.

Amastigotes: Intracellular developmental stages of *Leishmania* spp. and *Trypanosoma cruzi* in the vertebrate host.

Amplifier host: The host in which the organism pool is built without necessary causing an overt and severe disease.

Anthropophilic: A parasite that prefers human as a host to animals.

Antibiotic: A substance, produced by bacteria or fungi that destroys or prevents the growth of other microorganisms (bacteria and fungi).

Antibodies (immunoglobulins): The glycoproteins, present in the serum and body tissues that are induced when immunogenic molecules are introduced into the host's lymphoid system. Antibodies bind specifically to the antigen that induced their formation.

Antigen: Any substance capable of provoking the lymphoid tissues of an individual to respond by generating an immune reaction specifically directed at the inducing substance and not at other unrelated substances.

Antigenic drift: The antigenicities of the receptor-binding haemagglutinin and the neuraminidase are constantly changed under the selective pressure of immune responses. The change is based on point mutation. The antigenic drift is the most important mechanism for persistence of influenza A viruses in a population.

Antigenic shift: The change in antigenicity, which results from genetic reassortment. Antigenic shift is believed to be cause of pandemics in populations.

Antiseptic: An agent that destroys microorganisms, which is suitable for application to living tissues such as a wound.

Antiserum: The blood serum containing antibodies.

Antitoxin: A substance that specifically neutralizes the toxin.

Autogenous: Originating within an individual as transfer of tissue from one site to another in the same individual.

Autoinfection: The re-infection of a host by larvae of a parasite already within the host.

Bacteremia: The occurrence of bacteria transiently in the bloodstream.

Biosafety: Safety measures adopted in industries or laboratories dealing with biological materials or organisms.

Biosecurity: The measures that are taken to prevent the spread or introduction of harmful organisms to humans, animals and plant life.

Carrier: An individual that is infected with an infectious agent without manifesting clinical signs and that can be a source of infection to others. As a rule, carriers are less infectious than cases, but epidemiologically, they are more dangerous than cases because they escape recognition.

Case fatality rate: The number of deaths caused by a disease relative to the total number of diagnosed cases.

Case: An animal in a population identified as having a particular disease which may be subjected to epidemiological investigation.

Cercaria: A motile larval stage of digenean trematodes, infective for second intermediate host or final host.

Chemoprophylaxis: The prevention of disease by using chemical agents or drugs.

Chemotherapy: The treatment of disease by using chemical agents or drugs.

Chronic: Relating to a disease of slow progress and long duration.

Cleaning: A soil-removing process which removes many microorganisms. Cleaning is the necessary prerequisite for sterilization and disinfection.

Colonization: The establishment of a stable population of bacteria in the host.

Commensal microorganism: The normal flora lives on the skin and on the mucous membranes of the upper respiratory tract and intestine and obtains nourishment from the secretions and food residues.

Communicable disease: The disease capable of being directly or indirectly transmitted from human to human or animal to animal or from environment to human or animal.

Contagious disease: An infectious disease in which the causative agent(s) are spread by direct animal-to-animal (also human-to-human or animal-to-human) contact.

Contamination: The presence of an infectious agent on a body surface, also on or in clothes, beddings, surgical instruments or other inanimate articles or substances including water, milk and food.

Control of disease: The measures that is necessary to combat illness among members of a population and to eliminate cause of illness.

Dead-end host: A host from which infectious agents are not transmitted to other susceptible hosts, *e.g.*, humans are dead-end hosts for trichinellosis, because the larvae encysted in muscles and human flesh are unlikely to be a source of food for other animals susceptible to this parasite.

Definitive host: The final (specific) host in which the parasite completes its adult stage of life.

Dermatophytosis: A chronic fungal infection of the skin, hair or nails. This may be caused by zoophilic dermatophytes, which are transmitted to humans from overtly or latently infected animals.

Determinant: A factor that affects the health of a population.

Disinfectant: A chemical agent that is capable of disinfection. Disinfectant may kill vegetative bacteria, viruses, fungi and occasionally spores by the destruction of proteins, lipids or nucleic acids in the cell or its cytoplasmic membrane. Disinfectant is used on inanimate objects and surfaces.

Disinfection: The destruction of microorganisms, but not usually bacterial spores. Disinfection does not necessarily kill all microorganisms but reduce them to a level acceptable for a defined purpose.

Domiciliated animal: An animal lives in and around human dwellings.

Ecology: The study of the relationship between organisms and their environment.

Ecosystem: A community of living organisms (plants, animals and microbes) in conjunction with the nonliving components of their environment (air, water, soil *etc.*) interacting as system.

Elimination of disease: The process of removal of risk factors of the disease from the population with a view to reduce the prevalence of disease to reasonably low level. It is somewhat like a partial eradication.

Emerging zoonosis: A zoonosis that is newly recognized or newly evolved or that has occurred previously but shows an increase in incidence or expansion in geographical, host or vector range.

Endemic: The usual frequency of occurrence of a disease in a population. The constant presence of a disease in a population is also known as endemic occurrence of a disease. The term 'endemic' can be applied not only to overt disease but also to disease in the absence of clinical signs and to levels of circulating antibodies.

Endotoxin: A component of the outer membrane of Gram-negative bacteria, released from the bacterial surface following natural lysis of the bacterium or by disintegration of the organism *in vitro*.

Enterotoxin: A toxin which acts on the intestinal tract to cause excessive exudation of fluid.

Enzootic: Describing a disease of animals that is restricted to a locality or population.

Epidemic: The occurrence of an infectious or non-infectious disease to a level in excess of the expected (*i.e.*, endemic) level.

Epidemiology: The study of disease in populations and of factors that determine its occurrence. Veterinary epidemiology additionally includes investigation and assessment of other health-related events, notably productivity.

Epizootic: An occurrence of disease in excess of its anticipated frequency in animal population. It is synonymous with the term "epidemic" in human medicine.

Epornitics: The outbreak of disease in avian populations.

Eradication of disease: The process of complete removal of a disease from an area or country.

Etiology: The study of cause of a disease or abnormal condition.

Exotic disease: The disease which came from a foreign land.

Exotoxin: The diffusible proteins secreted into the external medium by the pathogen. Exotoxins are produced by Gram-negative and Gram-positive bacteria. Exotoxins are very toxic, highly antigenic and used for toxoid preparation.

Extrinsic incubation period: The time period between infection of a vector and attaining a stage at which it becomes able to transmit the infectious agent.

Exzootic disease: The disease which has been eradicated from a country.

Final host: The host in the course heterogenous development where sexual multiplication takes place.

Florid plaques: Accumulation of prions in intracellular vesicles in brain which increase into large vacuole. They are focal amyloid deposits surrounded by vacuolized cells.

Fomites: Inanimate or non-living objects that carry viable organisms.

Food chain: The simplistic view of the relationship between an animal and its food.

Four corners disease: A synonym of a disease "*Hantavirus* pulmonary syndrome" or "Sin Nombre disease.

Habitat: A place where an organism lives.

Hazards: Unacceptable contamination, growth of microorganisms, undesirable chemicals and physical agents or factors in food.

Health: A state of complete physical, mental and social well-being and not merely the absence of disease or infirmity.

Herd immunity: The level of resistance of a community or group of people to a particular disease.

Host: An organism (human, animal or plant) that harbours a parasite. Host may be of different types such as intermediate, definitive (final), dead-end, accidental (incidental) and amplifier hosts.

Hygiene: The conditions and practices that help to maintain health and prevent the spread of diseases.

Hyperendemic: The continuous presence of a disease to a high level in a population, affecting all age-groups equally.

Iatrogenic (physician-induced) disease: Any untoward or adverse consequence of a preventive, diagnostic or therapeutic regimen or procedure that causes impairment, handicap, disability or death resulting from a physician or professional activity.

Immunity: The ability of body to resist infection or invasion by other (foreign) agents or organisms and to counter any harmful effects caused by them.

Immunization: The process to develop active, passive or adoptive immunity in an individual by administration of vaccine, antiserum and lymphocytes, respectively.

Immunocompetent: An individual whose immune system works normally, capable to respond against an antigen.

Immunocompromised/immunosuppressed: An individual who does not have the ability to respond normally to an infection due to an impaired or weakened immune system. This inability to fight infection can be caused by a number of conditions including illness and diseases, *e.g.,* malnutrition, drugs, diabetes, HIV infection *etc.*

Incidence: The number of new cases that occur in a known population over a specified period of time. It is usually expressed in relation to the population at risk and the time during which the population is observed.

Incubation period: The time period between infection and occurrence of clinical symptoms.

Infection: The invasion of the body by infectious agents and their successful propagation in tissues and body fluid. Infection represents only the first stage *i.e.,* the entry of the agent into the body of the host, in a series of events that finally leads to the development of a disease. There are several levels of infections such as colonization, subclinical or inapparent infection, latent infection and manifest or clinical infection.

Infectious disease: The disease caused by an infectious agent.

Infectivity: The ability of an infectious agent to breach the new host's defences.

Infestation: The lodgment and development of parasites in or on the body of a host.

Intermediate host: The host in the course of heterogenous development in which asexual development (sometimes multiplication) takes place.

Intoxication: A type of food poisoning which arises due to consumption of food already containing microbial toxins in it (intradietetic or pre-formed toxins in the food).

Invasion: The entry of a pathogen across the epithelial cell surface (local invasion) or enter across the epithelial cell surface and spread systemically through the body.

Isolation: The separation for the period of communicability of infected person or animals from others in such places and under such conditions, as to prevent or limit the direct or indirect transmission of the infectious agent from those infected to those who are susceptible or who may spread the agent to others. Isolation is the oldest communicable disease preventive measure.

Latent infection: The infection in which the host does not shed the infectious agent which lies dormant within the host without symptoms and often without demonstrable presence in blood, tissue or body secretions of the host.

Management of disease: A set of activities aimed at controlling or eliminating the threat of a disease in an individual or population. The disease management can be done by prevention, control, elimination and eradication.

Monitoring: The routine collection of information on disease, productivity and other characteristics, possibly related to them in a population.

Morbidity: The amount of disease in a population. It is commonly defined in terms of incidence or prevalence.

Mortality: The measure of the number of deaths in a population. It represents the lot of animals dying or dead in a population.

Neglected zoonoses: The zoonoses that are not prioritized by health systems at national and international levels.

Niche: The functional position (role) of a species in a biotic community.

Nosocomial infection: An infection originating in a patient while in a hospital or other health care facilities.

Notifiable disease: The disease which requires immediate reporting to the higher health authority to take necessary action for preventing further spread.

Opportunistic infection: An infection by an organism that takes the opportunity provided by a defect in host immune system to infect the host and hence cause disease.

Opportunistic pathogens: Pathogens that rarely cause disease in individuals with intact immunological and anatomical defences. They cause disease only when the host defences are impaired or compromised, as a result of congenital or acquired disease or by the use of immunosuppressive therapy or surgical techniques.

Outbreak: An identified occurrence of disease involving one or more animals. In developed countries, an outbreak is frequently synonymous with disease occurrence on individual farms or holdings. Foodborne disease outbreak is an incident in which two or more persons experience a similar illness, usually gastrointestinal after ingestion of a common food, and epidemiological analysis implicates the food as the source of the illness.

Pandemic: A widespread epidemic that usually affects a large proportion of the population that may involve several countries or continents.

Parasite: An organism that lives in or on the body of a host.

Paratenic host: The host in which infective stages may accumulate but do not develop further.

Pasteurization: The process of heating of every particle of milk to at least 63 °C for 30 minutes, or 72 °C for 15 seconds, or any time-temperature combination which is equally efficient. The index organism for pasteurization is *Coxiella burnetii*.

Pathogen: An organism that parasitizes an animal, plant or human and produces a disease.

Pathogenesis: The process of developing a disease.

Pathogenicity: The capacity of a microbe to cause damage in a host.

Personal hygiene: The hygiene adopted by an individual.

Prepatent period: The time period between infection and first appearance of parasite or sexual parasite products like eggs, larvae *etc.*

Prevalence: The number of occurrences of disease, infection, antibody presence *etc.* in a population, usually relating to a particular point in time. It is commonly expressed as the proportion of the population at risk.

Prevention of disease: The measures that are taken to exclude a disease from an unaffected (healthy) population.

Primary pathogen: The pathogens that is capable to establish infection and cause disease in previously healthy individual with intact immunological defences. However, they may more readily cause disease in individuals with impaired defences.

Prion: A small virus like particle consisting of only protein (and no DNA or RNA).

Prophylaxis: The measures that are adopted to prevent the occurrence or spread of a disease.

Quarantine: A restraint, placed upon the movement of the animals, humans, plants or goods which are suspected of being carriers or vehicle of infection or of having been exposed to infection. In a general sense refers to detention of animals suspected of disease at ports or land borders.

Re-emerging zoonosis: An already known zoonotic disease that is either shifts its geographical settings or expands its host range or significantly increases its prevalence. Re-emergence of disease is mainly due to the international travel and trade.

Repellent: A substance applied on the skin, clothing or other places to keep away insects.

Reservoir: An animate or inanimate object on or in which an infectious agent usually lives and serve as a source of infection. Reservoirs are usually domestic or wild animals, birds and human beings that may or may not show clinical disease but help to maintain the infection for long periods.

Sanitation: The provision of facilities and services for the safe disposal of human urine and faeces. It also refers to the maintenance of hygienic conditions, through services such as garbage collection and wastewater disposal.

Sanitization: The process of destruction of all pathogenic and almost all non-pathogenic microorganisms from equipment surface, wounds or clothing, through the use of heat or chemicals.

Saprobe: An organism that depends on dead organic matters for its food and energy.

Sapronoses (saprozoonoses): The zoonoses which involve transmission of infectious agent to humans from environment but not directly from animals.

Sensitivity: Sensitivity of a diagnostic method is the proportion of true positives that are detected by the method. It is the ability of a test to correctly detect the presence of disease or infection.

Septicemia: An overwhelming invasion of the bloodstream from a focus of infection.

Source of infection: Any host organism, object or substance from which an infectious agent passes to a susceptible individual. A source of infection thus may be patient, carrier, reservoir, vector or vehicle.

Specificity: Specificity of the method is the proportion of true negatives that are detected. It is the ability of a test to correctly detect the non-diseased or healthy status of animal in herds.

Sporadic: A sporadic outbreak of disease is one that occurs irregularly and haphazardly.

Sterilization: The process used to achieve sterility, an absolute term meaning the absence of all viable microorganisms.

Surveillance: An epidemiological investigation involves gathering, recording and analysis of data; and dissemination of information to interested parties, so that action can be taken to control disease.

Survey: An epidemiological investigation involves collection of information in which a causal hypothesis usually is not tested.

Synanthropic: One that has a commensal relationship with humans.

Titre: The extent to which a solution of an antibody can be diluted before it ceases to cause agglutination of the matching antigen in vitro.

Toxemia: A condition of the presence of microbial or endogenous toxins in the blood circulation.

Toxigenicity: The toxin producing property of the organism.

Toxin: A poisonous substance, usually of biological origin and especially produced by a bacterium. Toxin is protein in nature and capable of inducing antibody production.

Toxoid: A detoxified exotoxin of the microorganisms which is used in the preparation of vaccines. In general, toxoid preparations are highly efficacious and safe immunizing agent.

Transmission: The passing of a pathogen causing communicable disease from an infected host or group of a particular individual or group, regardless of whether the other individual was previously infected.

Vaccine: An 'immuno-biological substance' designed to produce specific protection against a given disease. It stimulates the production of protective antibody and other immune mechanisms. Vaccines may be prepared from live modified organisms, inactivated or killed organisms, extracted cellular fractions, toxoid or combination of these.

Vector: An invertebrate animal (frequently an arthropod) responsible for transmission of an infectious agent from an infected individual or its excreta to susceptible individual (or to some immediate source of infection such as food or water). Transmission of an infectious agent by a vector may be biological or mechanical.

Vehicle: Any nonliving thing or substance including food, milk, meat, water, pus, serum *etc.* by which or upon which an infectious agent passes from an infected individual to a susceptible one.

Virulence: The relative capacity of a microbe to cause damage in a host.

Zoonoses: Those diseases and infections that are naturally transmitted between vertebrate animals and humans.

Zoophilic: A parasite preferring animals to humans as host.

Appendix II

Zoonoses Involving Various Animal Species

There are various animal species that play their important role in the maintenance and transmission of particular zoonotic diseases. Some important zoonoses involving various animal species are as follows:

Cattle

Anthrax, brucellosis, campylobacteriosis, enterohemorrhagic *E. coli* infections, leptospirosis, listeriosis, Lyme borreliosis, Q fever, salmonellosis, staphylococcosis, tuberculosis, yersiniosis, bovine spongiform encephalopathy, cowpox, Crimean-Congo hemorrhagic fever, foot and mouth disease, pseudocowpox, rabies, Rift Valley fever, Russian spring summer encephalitis, vesicular stomatitis, African trypanosomiasis, cryptosporidiosis, giardiasis, fascioliasis, schistosomiasis, echinococcosis, taeniasis *etc.*

Sheep and Goats

Anthrax, brucellosis, campylobacteriosis, chlamydiosis, erysipeloid, leptospirosis, listeriosis, melioidosis, pasteurellosis, Q fever, salmonellosis, tuberculosis, tularemia, yersiniosis, Central European encephalitis, Crimean-Congo hemorrhagic fever, foot and mouth disease, louping ill, rabies, Rift Valley fever, Russian spring summer encephalitis, cryptosporidiosis, giardiasis, toxoplasmosis, fascioliasis, echinococcosis *etc.*

Canines

Brucellosis, ehrlichiosis, leptospirosis, Lyme borreliosis, Rocky Mountain spotted fever, Mediterranean spotted fever, Q fever, rabies, giardiasis, cryptosporidiosis, Chagas' disease, visceral larva migrans, cutaneous larva migrans, leishmaniasis, scabies, paragonimiasis, clonorchiasis, echinococcosis, dipylidiosis, *Brugia* filariasis, dirofilariasis *etc.*

Felines

Cat scratch disease, murine typhus, Q fever, rabies, cryptosporidiosis, giardiasis, toxoplasmosis, American trypanosomiasis, visceral larva migrans, cutaneous larva migrans, clonorchiasis, paragonimiasis, echinococcosis *etc.*

Equines

Anthrax, Lyme borreliosis, leptospirosis, melioidosis, salmonellosis, yersiniosis, American equine encephalitis, eastern equine encephalitis, western equine encephalitis, Venezuelan equine encephalitis, equine influenza, rabies, Semliki forest fever, West Nile fever, vesicular stomatitis *etc.*

Swine

Brucellosis, campylobacteriosis, erysipeloid, leptospirosis, salmonellosis, yersiniosis, foot and mouth disease, Japanese encephalitis, swine influenza, swine vesicular disease, cryptosporidiosis, toxoplasmosis, taeniasis, trichinellosis *etc.*

Rodents

Leptospirosis, Lyme borreliosis, Mediterranean spotted fever, murine typhus, plague, rat bite fever, relapsing fever, Q fever, Rocky Mountain spotted fever, scrub typhus, tuberculosis, yersiniosis, Argentine hemorrhagic fever, Bolivian hemorrhagic fever, California encephalitis, Colorado tick fever, Crimean-Congo hemorrhagic fever, *Hantavirus* infections, rickettsial pox, trypanosomiasis, visceral leishmaniasis, visceral larva migrans, trichinellosis *etc.*

Bats

Hendra virus hemorrhagic fever, *Nipah* virus encephalitis, rabies, Venezuelan equine encephalitis *etc.*

Monkeys

Tuberculosis, chikungunya fever, dengue fever, *Ebola* virus hemorrhagic fever, Kyasanur forest disease, *Marburg* virus hemorrhagic fever, Mayaro fever, monkeypox, rabies, yellow fever, simian malaria *etc.*

Birds

Campylobacteriosis, erysipeloid, pasteurellosis, Q fever, salmonellosis, yersiniosis, avian influenza, Central European encephalitis, chikungunya fever, Crimean-Congo hemorrhagic fever, Eastern equine encephalitis, Kyasanur forest disease, Japanese encephalitis, louping ill, Mayaro fever, Murray Valley encephalitis, Newcastle disease, Rocio encephalitis, Russian spring summer encephalitis, Sindbis fever, St. Louis encephalitis, Venezuelan equine encephalitis, West Nile fever, Western equine encephalitis *etc.*

Wildlife (Other than monkeys and birds)

Anthrax, brucellosis, campylobacteriosis, erysipeloid, leptospirosis, listeriosis, Lyme borreliosis, pasteurellosis, plague, Q fever, rat bite fever, rickettsialpox, salmonellosis, tuberculosis, tularemia, yersiniosis, Central European encephalitis, elephantpox, foot and mouth disease, Kyasanur forest disease, louping ill, lymphocytic choriomeningitis, rabies, Russian spring summer encephalitis, cryptosporidiosis, toxoplasmosis, clonorchiasis, dicrocoeliasis, fascioliasis, paragonimiasis, echinococcosis, dirofilariasis, trichinellosis *etc.*

Amphibians, Snakes and Turtles

Listeriosis, salmonellosis, dioctophymiasis, gnathostomiasis, pentastomidiasis, sparganosis *etc.*

Fish and Crustaceans

Cholera, erysipeloid, tuberculosis, vibrioses, clonorchiasis, echinostomiasis, opisthorchiasis, paragonimiasis, diphyllobothriasis, sparganosis, anisakiasis, acanthocephaliasis *etc.*

Appendix III

Table 1: Infections Resulting from Animal Bite

Animal species	Infections Due to different Organisms Resulting from Animal Bite
Dog	*Actinomyces, Bacillus, Bacteroides, Citrobacter, Clostridium, Corynebacterium, Enterococcus, Escherichia, Fusobacterium, Haemophilus, Lactobacillus, Moraxella, Neisseria, Pasteurella, Proteus, Pseudomonas, Staphylococcus, Streptococcus* spp. etc.
Cat	*Acinetobacter, Actinomyces, Aeromonas, Alcaligenes, Bacillus, Bacteroides, Clostridium, Enterococcus, Klebsiella, Pasteurella, Pseudomonas, Sporothrix, Staphylococcus, Streptococcus, Yersinia* spp. etc.
Bat	Rabies virus (mainly *Lyssavirus* serotype 1).
Simian	*Bacteroides, Clostridium, Enterococcus, Haemophilus, Neisseria, Staphylococcus, Streptococcus* spp. etc.
Sheep	*Actinobacillus, Pasteurella* spp. etc.
Camel	*Bacillus, Escherichia, Klebsiella, Pseudomonas, Staphylococcus, Streptococcus* spp. etc.
Horse	*Actinobacillus, Clostridium, Escherichia, Neisseria, Staphylococcus, Streptococcus* spp. etc.
Pig	*Bacteroides, Escherichia, Proteus, Staphylococcus, Streptococcus* spp. etc.
Rat	*Corynebacterium, Fusobacterium, Leptospira, Pasteurella, Spirillum, Staphylococcus, Streptobacillus* spp. etc.
Squirrel	*Corynebacterium, Francisella, Staphylococcus, Streptococcus* spp. etc.
Hamster	*Acinetobacter, Leptospira* spp., lymphocytic choriomeningitis virus. etc.
Opossum	*Aeromonas, Citrobacter, Escherichia, Pasteurella, Staphylococcus, Streptococcus* spp. etc.
Snake	*Acinetobacter, Alcaligenes, Bacillus, Bacteroides, Citrobacter, Clostridium, Corynebacterium, Enterobacter, Micrococcus, Proteus, Pseudomonas, Salmonella, Staphylococcus, Streptococcus* spp. etc.
Lizard	*Serratia, Staphylococcus* spp. etc.
Alligator	*Acinetobacter, Aeromonas, Alcaligenes, Bacillus, Bacteroides, Citrobacter, Clostridium, Corynebacterium, Enterobacter, Micrococcus, Proteus, Pseudomonas, Salmonella, Staphylococcus, Streptococcus* spp. etc.
Shark	*Aeromonas, Citrobacter, Micrococcus, Pseudomonas, Staphylococcus, Vibrio* spp. etc.
Bird	*Aspergillus, Bacteroides, Clostridium, Pseudomonas, Streptococcus* spp. etc.

Appendix IV

Table 2: Principal Features of Important Zoonotic Diseases

Disease	Causative Agent	Geographic Occurrence	Transmission	Characteristics in Humans	Management in Humans
Bacterial zoonoses					
Anthrax	*Bacillus anthracis*	Worldwide	Contact of abraded skin with infected hide or fur of animals. Inhalation of contaminated wool or dusts. Ingestion of infected milk/meat.	Carbuncles on skin, dyspnea, abdominal pain, meningitis *etc.*	Hygienic measures. Proper disposal of unopened animal carcases. Disinfection of animal products, wool, hair *etc.* Vaccination with attenuated live spore vaccine or aluminium hydroxide adsorb cell free filtrate. Treatment with penicillin G.
Bartonelloses: Cat scratch disease, trench fever and Carrion's disease					
Cat scratch disease	*Bartonella henselae*	Worldwide	Cat scratch or bite (inoculation of infected cat flea faeces at the time of injury).	Lymphadenopathy, papule on skin *etc.*	Avoid contact with flea infested cats. Treatment with azithromycin.
Trench fever	*Bartonella quintana*	Worldwide	Human body louse (*Pediculus humanus*)	Fever, headache, rash, papule on skin, lymphadenopathy *etc.*	Personal hygiene. Avoid exposure to human body lice.
Carrion's disease	*Bartonella bacilliformis*	America	*Sandfly (Lutzomyia verrucarum) bites.*	Fever, headache, myalgia, abdominal pain, anemia, nodular growths under skin, vascular lesions *etc.*	Control of sandflies by using insecticides, repellents and protective clothing. Treatment with penicillin, streptomycin, tetracycline or chloramphenicol.

Disease	Causative Agent	Geographic Occurrence	Transmission	Characteristics in Humans	Management in Humans
Borrelioses: Lyme borreliosis and Relapsing fever					
Lyme borreliosis	*Borrelia burgdorferi*	Worldwide	Tick (*Ixodes* spp.) bite.	Erythema migrans on skin, regional lymphadenopathy *etc.*	Protection from tick. Prophylaxis with penicillin. Treatment with doxycycline.
Relapsing fever	*Borrelia recurrentis* and *B. duttonii.*	Worldwide	Human body louse (*Pediculus humanus corporis*) and tick (*Ornithodoros* spp.) bites.	Fever, nausea, vomiting, splenomegaly, hepatomegaly *etc.*	Control of lice and ticks. Protection from lice and ticks. Treatment with doxycycline or erythromycin.
Brucellosis	*Brucella melitensis, B. abortus, B. suis* and *B. canis.*	Mainly in Mediterranean countries but also in Asia, Africa and Latin America	Contact with excretions of infected animals. Ingestion of infected milk. Inhalation of contaminated dusts.	Lymphadenopathy, hepatomegaly, orchitis, epididymitis, prostitis *etc.*	Elimination of *Brucella*-infected animals. Laboratory precautions. Personal hygiene. Vaccination with *B. abortus* strains 19 BA and 104 M. Treatment with doxycycline plus streptomycin. Heat treatment of milk and milk products.
Chlamydial infections (Psittacosis)	*Chlamydophila psittaci, C. abortus* and *C. felis*	Worldwide	Inhalation of contaminated dust. Contact with excretions of infected animals.	Pneumonia, photophobia, fever *etc.*	Use of protective clothing and face masks by personnel. Treatment with doxycycline or azithromycin.
Ehrlichiosis	*Ehrlichia canis*	Mainly in US and UK	Ticks (*Amblyomma, Dermacentor* and *Ixodes* spp.) bites.	Pancytopenia, myalgia, arthralgia *etc.*	Use of protective clothing and insect repellents. Treatment with doxycycline.
Erysipeloid	*Erysipelothrix rhusiopathiae*	Worldwide	Contact of injured skin with infected animal tissues and contaminated instruments.	Skin lesions (extensive redness, edema and hemorrhages).	Avoid contact with infected animals. Wear the gloves for handling of contaminated materials. Treatment with penicillin or erythromycin.
Glanders	*Burkholderia mallei*	Asia	Contact of skin or mucous membranes with nasal secretions, ulcerating lesions, pus and mucous of infected animals. Inhalation of contaminated dusts.	Pustular skin lesions, necrosis in respiratory tract, pneumonia *etc.*	Culling and safe disposal of infected carcase. Use of protective clothing and gloves. Surgical drainage of abscesses. Treatment with ceftazidime or cotrimoxazole.

Disease	Causative Agent	Geographic Occurrence	Transmission	Characteristics in Humans	Management in Humans
Leptospirosis	*Leptospira interrogans*	Worldwide	Contact of skin and mucous membranes with urine of infected animals or water contaminated with it. Bites by infected rats, mice or hamsters (voiding of urine at the time of bite).	Fever, cough, conjunctival hyperemia, hepatomegaly, jaundice *etc.*	Control of rats and mice. Avoid swimming in contaminated water. Vaccination in farm animals. Treatment with penicillin G.
Listeriosis	*Listeria monocytogenes*	Worldwide	Contact with diseased animals during parturition. Inhalation of contaminated dusts. Ingestion of infected milk and milk products, pork and poultry meat.	Meningitis, encephalitis *etc.*	Heat treatment of milk and milk products, meat and vegetable. Treatment with ampicillin.
Melioidosis	*Burkholderia pseudomallei*	Southeast Asia	Contact of abraded skin with contaminated soil. Ingestion of infected pork, horse meat, milk and milk products from sheep/ goats. Inhalation of contaminated air.	Abscess formation in the organs, meningoencephalitis *etc.*	Avoid contact with contaminated soil and surface water. Treatment with ceftazidime.
Pasteurellosis	*Pasteurella multocida*	Worldwide	Bite and scratches by infected animals (cats, dogs, rabbits, guinea pigs *etc.*). Rarely by ingestion and inhalation.	Redness and painful swelling of bite/scratch wound *etc.*	Hygienic measures. Treatment with penicillin or doxycycline.
Plague	*Yersinia pestis*	Asia, Africa and US	Infected flea (*Xenopsylla* spp.) bite. Inhalation of aerosols from patients of pneumonic plague.	Lymphadenopathy, pneumonia, septicemia *etc.*	Hygienic measures. Control of rats and fleas. Treatment with streptomycin.
Rat bite fever	*Spirillum minus* and *Streptobacillus moniliformis*	Worldwide	Infected rat bite (transmission occurs through blood and saliva of infected rat).	Pain, edema and ulceration on bite wound, fever, regional lymphadenopathy *etc.*	Control of rats. Management of wound. Treatment with penicillin.

Rickettsioses: Rocky Mountain spotted fever (RMSF), Mediterranean spotted fever (MSF), Epidemic typhus, Murine typhus, Indian tick typhus, Rickettsial pox, Scrub typhus and Q fever

Disease	Causative Agent	Geographic Occurrence	Transmission	Characteristics in Humans	Management in Humans
Rocky Mountain spotted fever	*Rickettsia rickettsii*	America	Infected tick (*Amblyomma*, *Dermacentor* and *Rhipicephalus* spp.) bite.	Fever, maculopapular rash, hemorrhages *etc.*	Control of ticks. Use of protective clothing and tick repellents. Avoid exposure to natural foci. Treatment with doxycycline.
Mediterranean spotted fever	*Rickettsia conorii*	Europe, Asia and Africa	Infected tick (*Rhipicephalus sanguineus*) bites.	Necrosis on the site of tick bite, regional lymphadenitis, maculopapular rash *etc.*	Control of ticks. Use of tick repellents and protective clothing. Treatment with doxycycline.
Epidemic typhus	*Rickettsia prowazekii*	Africa and America	Bite by human body louse (contact of its faeces during bite). Airborne infection with its dried faeces.	Pink spotted maculopapular rash on skin, encephalitis, myocarditis *etc.*	Control of louse. Prophylaxis and treatment with doxycycline. Vaccination with attenuated live virus vaccine.
Murine typhus	*Rickettsia typhi*	UK and US	Rat fleas (*X. cheopis* and *Leptopsylla segnis*) bites.	Fever, rash, epistaxis *etc.*	Control of rats and rat fleas. Treatment with doxycycline.
Indian tick typhus	*Rickettsia conorii* subsp. *indica*	India (Maharashtra, Punjab, Karnataka, Kerala, U.P. M.P., H.P. and J & K)	Infected tick (*Boophilus*, *Haemaphysalis*, *Ixodes* and *Rhipicephalus*) bites.	Fever, severe headache and maculopapular rash which often turns purpuric.	Control of ticks. Personal protection from ticks. Treatment with doxycycline.
Rickettsial pox	*Rickettsia akari*	US and former USSR	Infected mite (*Liponyssoides sanguineus*) bites.	Papule, enlargement of regional lymph nodes *etc.*	Control of mites and mice. Use of mite repellents and protective clothing. Treatment with doxycycline.
Scrub typhus	*Orientia tsutsugamushi*	Southeast Asia	Infected mite (*Trombicula* spp.) bites. Direct contact with infected mites.	Papule, generalized lymphadenopathy *etc.*	Control of mites and mice. Use of protective clothing and mite repellants. Treatment with doxycycline.

Disease	Causative Agent	Geographic Occurrence	Transmission	Characteristics in Humans	Management in Humans
Q fever	*Coxiella burnetii*	Worldwide	In humans-Contact with infected animals or their dried excreta. Ingestion of infected milk and milk products. Inhalation of contaminated dust.	Fever, chills, headache, myalgia, arthralgia *etc.*	Hygienic measures. Heat treatment of milk and milk products. Control of ticks in livestock raising areas. Vaccination with Q-vax vaccine. Treatment with doxycycline or tetracycline.
Staphylococcal infections	*Staphylococcus aureus* and *S. intermedius.*	Worldwide	Contact of skin with infected tissues and skin lesions of animals. Bites and scratches by infected animals.	Abscess, folliculitis, impetigo, generalized pyoderma *etc.*	Hygienic measures. Avoid animal bites and scratches. Wear the gloves when working with infected animals tissues. Treatment with penicillin.
Streptococcus equi infections	*Streptococcus equi* subsp. *zooepidemicus*	UK	Contact of skin with nasal secretions of infected animals. Bites by infected animals. Ingestion of infected milk and milk products.	Impetigo, pneumonia, septicemia, meningitis, lymphadenopathy *etc.*	Avoid contact with infected animals. Heat treatment of milk and milk products. Treatment with penicillin G.
Streptococcus suis infections	*Streptococcus suis* serovars.	UK and North America	Contact of skin and conjunctiva with infected pigs. Contact of wound with contaminated instruments. Ingestion of infected pork.	Meningitis, loss of hearing and balance, deafness *etc.*	Wear the gloves to avoid contact with infected pigs. Heat treatment of pork. Prophylaxis and treatment with penicillin G.
Tuberculosis	*Mycobacterium tuberculosis, M. bovis, M. africanum, M. microti* and *M. canettii.*	Worldwide	Inhalation of infected aerosols and contaminated dusts. Ingestion of infected milk and meat. Contact of injured skin and mucous membranes with infected animal tissues.	Continuous fever, persistent cough, chest pain, hemoptysis *etc.*	Hygienic measures. Use of face masks by patients and persons caring them. Heat treatment of milk. BCG vaccination. Treatment with rafampicin, isoniazid, pyrazinamide and ethambutol.
Tularemia	*Francisella tularensis* and *F. philomiragia*	US, Europe and Russia	Contact with excreta, blood or organs of infected wild animals. Bites by ticks, lice, fleas and flies following bites of infected cats and squirrels. Inhalation of dusts contaminated with rodent excreta. Ingestion of contaminated food and water.	Cutaneous ulcer, septicemia, lymphadenopathy *etc.*	Take precautions when in contact with wild animals, especially rodents. Vaccination with attenuated strains of *F. tularensis.* Treatment with streptomycin or ciprofloxacin.

Foodborne bacterial diseases: *Aeromonas hydrophila* infection, *Bacillus cereus* food poisoning, Botulism, *Clostridium perfringens* gastroenteritis, Campylobacteriosis, *Escherichia coli* infection, Salmonellosis, Staphylococcal food intoxication, Vibrioses and Yersinioses.

Disease	Causative Agent	Geographic Occurrence	Transmission	Characteristics in Humans	Management in Humans
Aeromonas hydrophila infection	*Aeromonas hydrophila*	Worldwide	Ingestion of contaminated seafood, chicken, red meat and water.	Gastroenteritis, diarrhea *etc.*	Prevention of contamination of foods and water. Adequate heat treatment of foods.
Bacillus cereus food poisoning	*Bacillus cereus*	Worldwide	Ingestion of contaminated boiled rice (dried and stored), meat and vegetables.	Nausea, vomiting, diarrhea, abdominal pain *etc.*	Hygienic measures. Storage of food at refrigeration temperature. Symptomatic treatment.
Botulism	*Clostridium botulinum*	Worldwide	Ingestion of foods containing preformed toxins.	Dysphagia, quadriplegia, dysarthria, diplopia *etc.*	Active immunization using botulinum toxoid. Neutralization of toxin using antitoxin.
Campylobacteriosis	*Campylobacter jejuni, C. coli* and *C. lari*	Worldwide	Ingestion of contaminated poultry meat, pork, milk and water. Contact with secretions of infected animals.	Acute enteritis, diarrhea, colic *etc.*	Hygienic handling of food. Heat treatment of milk, meat and their products. Fluid and electrolytes therapy. Treatment with doxycycline.
Clostridium perfringens gastroenteritis	*Clostridium perfringens*	Worldwide	Ingestion of contaminated meat and meat products.	Diarrhea, abdominal cramps *etc.*	Maintain personal hygiene. Thorough cooking of foods. Refrigerated storage of foods.
Escherichia coli infection	*Escherichia coli*	Worldwide	Ingestion of contaminated ground meat products, hamburgers, salami, milk and milk products.	Fever, vomiting, diarrhea, hemorrhagic colitis, haemolytic uremic syndrome *etc.*	Hygienic measures. Prevention of contamination of food and water. Heat treatment of foods. Health education of public.
Salmonellosis	*Salmonella serovars*	Worldwide	Ingestion and contact of contaminated foodstuffs (poultry meat, pork, milk and milk products, eggs and egg products).	Fever, nausea, vomiting, diarrhea *etc.*	Hygienic measures. Prohibition of infected and carrier persons to work in any food-processing and food-serving establishment. Heat treatment of milk and meat. Treatment with ciprofloxacin.
Staphylococcal food intoxication	*Staphylococcus aureus*	Worldwide	Ingestion of toxins pre-formed in the food.	Vomiting, abdominal cramps, diarrhea *etc.*	Prevention of contamination of food. Heat treatment of foods like milk, meat, poultry sausages, fish *etc.*

Disease	Causative Agent	Geographic Occurrence	Transmission	Characteristics in Humans	Management in Humans
Vibrio cholerae infection (Cholera)	*Vibrio cholerae*	Worldwide	Ingestion of infected raw fish, crabs, seafood and oysters.	Watery diarrhea, vomiting *etc.*	Fluid and electrolyte therapy. Treatment with ciprofloxacin or doxycycline.
Vibrio parahaemolyticus infection	*Vibrio parahaemolyticus*	Worldwide	Ingestion of raw or insufficiently cooked seafood like shellfish.	Diarrhea, vomiting, fever *etc.*	Hygienic measures. Adequate cooking of seafood. Fluid and electrolyte therapy.
Yersinioses	*Yersinia enterocolitica* and *Y. pseudo-tuberculosis*	Worldwide	Ingestion of infected milk and pork.	Enterocolitis, mesenteric and ileocecal lymphadenitis *etc.*	Avoid contact with infected animals. Follow the food hygiene. Treatment with ciprofloxacin or doxycycline.

Viral zoonoses

Zoonoses caused by Rhabdoviruses: Rabies and Vesicular stomatitis.

Disease	Causative Agent	Geographic Occurrence	Transmission	Characteristics in Humans	Management in Humans
Rabies	*Lyssavirus* type 1	Worldwide	Infected animal bites and licks. Inhalation of infected aerosols.	Nervous excitement, paralysis of muscles *etc.*	Control of stray dogs. Management of bite wound. Vaccination of owned dogs and persons at risk. Symptomatic treatment.
Vesicular stomatitis	*Vesiculovirus*	America, Africa and Asia	Sandfly (*Phlebotomus* spp.) bite.	Fever, headache, myalgia, arthralgia *etc.*	Hygienic measures. Vaccination with formalin-inactivated and attenuated live virus vaccine.

Zoonoses caused by Prions: Bovine spongiform encephalopathy and Variant Creutzfeldt-Jakob disease

Disease	Causative Agent	Geographic Occurrence	Transmission	Characteristics in Humans	Management in Humans
Bovine spongiform encephalopathy	Prion	Europe	In cattle- Consumption of bovine meat meal and bone meal contaminated with prion.	Degenerative changes in bovine brain tissues.	Disease notification. Strict hygienic measures. Ban on bovine meat meal and bone meal from BSE outbreak areas. Slaughter of infected bovines and hygienic disposal. Sterilization of ruminant derived feeds.
Variant Creutzfeldt Jakob disease	Prion	Europe, Asia and America	In humans-consumption of meat and offal contaminated with prion.	Depression, anxiety, difficulty in walking *etc.*	Prevention on entry of infectious agent in the animal or human food chain. Ban on the meat and offal from the disease outbreak areas.

Disease	Causative Agent	Geographic Occurrence	Transmission	Characteristics in Humans	Management in Humans
Zoonoses caused by Orthomyxoviruses: Avian influenza and Swine influenza					
Avian influenza	Influenza virus A(H5N1) and A(H7N9)	Worldwide	Contact with infected live birds.	Fever, splenomegaly, icterus etc.	Strict hygienic measures. Slaughter of all birds of a poultry farm having disease outbreak. Treatment with amantadine, rimantadine, zanamivir or oseltamivir.
Swine influenza	Influenza virus A(H1N1)	Worldwide	Inhalation of infected aerosols. Contact with infected person.	Fever, bronchitis, pneumonia, myalgia etc.	Strict hygienic measures. Treatment with amantadine, rimantadine, zanamivir or oseltamivir.
Zoonoses caused by Coronavirus: Severe acute respiratory syndrome (SARS)					
Severe acute respiratory syndrome	SARS Coronavirus	Worldwide	Inhalation of infected aerosols. Contact with infected person.	Fever, cough, myalgia etc.	Strict hygienic measures. Use of face mask. Symptomatic treatment.
Zoonoses caused by Poxviruses: Contagious ecthyma of sheep (ORF), Milker's nodules (Pseudocowpox), Cowpox, Buffalopox and Monkeypox					
Contagious ecthyma of sheep (Orf)	Orf virus (a member of the genus *Parapoxvirus*)	Worldwide	In sheep and goats, the virus enters through abrasions on the skin of lips and face. In humans, through direct contact with an infected animal or indirectly from a contaminated environment.	Lesions on the hands and arms, fever and axillary lymphadenitis.	Avoid contact of abraded or cut skin with infected animals, scabs and crusts, wool or hides. Treatment is symptomatic and supportive.
Milker's nodules (Pseudo-cowpox)	Pseudocowpox virus (genus *Parapoxvirus*)	Europe and US	Contact with pox lesions on the udder of milking cows.	Vesicles on skin, inflammation of axillary lymph nodes etc.	Avoid contact with pox lesions on the udder of infected cows.
Cowpox	Cowpox virus (genus *Orthopoxvirus*)	Asia, Europe and US	Contact with pox lesions on the udder of cows.	Pox lesions on hands and face, conjunctivitis, lymphangitis etc.	Vaccination with vaccinia virus (protection from orthopoxviruses). Treatment with cidofovir.
Buffalopox	Buffalopox virus (genus *Orthopoxvirus*)	Asia	Contact with pox lesions on the udder of buffalo.	Pox-like skin eruptions on hands etc.	Vaccination with vaccinia virus (protection for orthopoxviruses). Treatment with cidofovir.

Disease	Causative Agent	Geographic Occurrence	Transmission	Characteristics in Humans	Management in Humans
Monkeypox	Monkeypox virus (genus *Orthopoxvirus*)	Africa and US	Unknown	Skin eruptions, lymphadenopathy *etc.*	Vaccination with vaccinia virus (protection for orthopoxviruses). Treatment with cidofovir.
Zoonoses caused by Arenaviruses: Lassa fever and Lymphocytic choriomeningitis (LCM)					
Lassa fever	*Lassa* virus	West Africa	Ingestion of food contaminated with faeces of infected rodent (mice of the genus *Mastomys*). Consumption of killed infected mice.	High fever, headache, myalgia, cervical lymphadenopathy *etc.*	Control of rodents in human dwellings. Prevention of contamination of foods from rats. Isolation of patients in hospitals to prevent the spread in others. Treatment with ribavirin.
Lymphocytic chorio-meningitis (LCM)	LCM virus	Worldwide	In humans, through contact with infected laboratory animals (hamster and mice), and also through bites and aerosols. Ingestion of foods contaminated with excretions of rodents.	Fever, headache, photophobia, sneezing, bronchitis, meningoencephalitis or encephalitis *etc.*	Control of rodents. Careful handling of laboratory mice and products (production and use of monoclonal antibodies) derived from them. Treatment with ribavirin.
Zoonoses caused by Paramyxoviruses : Nipah virus encephalitis and Newcastle disease					
Nipah virus encephalitis	Nipah virus (genus *Henipavirus* and family Paramyxo-viridae)	Malaysia	In pigs, through contact of urine of infected bats. In humans, through direct contact with infected pigs.	Fever, headache, nausea, vomiting, symptoms related to encephalitis *etc.*	Prevention of pigs from contact with urine of flying foxes. Symptomatic treatment.
Newcastle disease	Newcastle disease virus (genus *Rubulavirus* and family Paramyxo-viridae)	Worldwide	In humans, through air or smear infection with conjunctiva and mucosa after direct contact with infected fowl.	Haemorrhagic conjunctivitis, fever, headache, rigor *etc.*	Personal hygiene. Use of goggles and face mask while handling infected poultry or live virus vaccine. Vaccination in poultry. Symptomatic treatment.

Disease	Causative Agent	Geographic Occurrence	Transmission	Characteristics in Humans	Management in Humans
Zoonoses caused by Reoviruses: Colorado tick fever					
Colorado tick fever	Colorado tick fever virus (genus *Coltivirus*, and family Coltiviridae)	Colorado, Rocky Mountains and western province of Canada	Infected tick (*Dermacentor andersoni*) bites. Human-to-human transmission through blood transfusion.	Fever, headache, myalgia, conjunctivitis, p h o t o p h o b i a , retrobulbar pain *etc.*	Use of protective clothing and repellents in endemic areas. Avoid blood donation by infected person. Symptomatic treatment.
Zoonoses caused by Alphaviruses: Chikungunya fever, Eastern equine encephalitis (EEE), Western equine encephalitis (WEE), Venezuelan equine encephalitis (VEE), Semliki forest fever and Sindbis fever.					
Chikungunya fever	*Alphavirus*	Africa and Southeast Asia	Mosquitoes (*Aedes aegypti, A. africanus* and *Culex* spp.) bites.	A r t h r a l g i a , maculopapular rash *etc.*	Control of mosquitoes. Use of mosquito nets and repellents. Vaccination with attenuated live virus vaccine.
Eastern equine encephalitis	*Alphavirus*	America and Asia	Mosquitoes (*Aedes sollicitans, A. vexans* and *A. albopictus*) bites.	Fever, encephalitis, nausea, vomiting *etc.*	Control of mosquitoes. Protection from mosquitoes. Vaccination of laboratory personnel and horses with formalin-inactivated vaccine.
Western equine encephalitis	*Alphavirus*	America	Mosquito (*Culex tarsalis*) bite.	Fever, encephalitis, myalgia *etc.*	Control of mosquitoes. Use of mosquito nets and repellents. Vaccination of laboratory personnel and horses with formalin-inactivated vaccine.
Venezuelan equine encephalitis	*Alphavirus*	America	Mosquito (*Culex* spp.) bite.	Viraemia, high fever, encephalitis *etc.*	Control of mosquitoes. Vaccination with formalin-inactivated vaccine.
Semliki forest fever	*Alphavirus*	Africa and Asia	Mosquito (*Aedes abnormalis*) bites. Inhalation of contaminated aerosols.	Fever, myalgia, arthralgia, headache *etc.*	Follow the regulations for protection and laboratory hygiene.
Sindbis fever	*Alphavirus*	Africa and Europe	Mosquitoes (*Anopheles, Aedes, Mansonia* and *Culex* spp.) bites.	Fever, arthralgia *etc.*	Control of mosquitoes.

Zoonoses caused by Flaviviruses: Dengue fever, Japanese encephalitis, Murray valley encephalitis, West Nile fever, Yellow fever, Kyasanur forest disease (KFD), Omsk hemorrhagic fever and Tick-borne encephalitis (Central European encephalitis-CEE and Russian spring summer encephalitis-RSSE)

Disease	Causative Agent	Geographic Occurrence	Transmission	Characteristics in Humans	Management in Humans
Dengue fever	*Flavivirus*	Asia, Africa and America	Mosquitoes (*Aedes aegypti, A. albopictus, A. scutellaris* and *A. africanus*) bites.	Fever, hemorrhagic diathesis, macular rash, hepatomegaly, shock *etc.*	Control of mosquitoes. Blood transfusion, replacement of fluids and plasma proteins and management of shock. Vaccination with tetravalent attenuated live virus vaccine.
Japanese encephalitis	*Flavivirus*	Asia	Mosquito (*Culex* and *Aedes* spp.) bite.	Fever, encephalitis, headache *etc.*	Control of mosquitoes. Rice cultivation and pig farming in separate locations. Vaccination with killed mouse brain vaccine.
Murray Valley encephalitis	*Flavivirus*	Australia	Mosquito (*Culex annulirostris*) bite.	Fever, encephalitis, rash *etc.*	Control of mosquitoes. Vaccination with JE virus vaccine (protective effect for Murray Valley encephalitis).
West Nile fever	*Flavivirus*	Africa and America	Mosquitoes (*Culex univittatus* and *Aedes japonicus*) bites.	Fever, myalgia, pharyngitis, rash *etc.*	Control of mosquitoes.
Yellow fever	*Flavivirus*	Africa and America	Mosquitoes (*Aedes aegypti* and *A. simpsoni*) and tick (*Haemagogus* spp.) bites.	Fever, icterus *etc.*	Control of mosquitoes and ticks. Vaccination with attenuated live virus mouse brain vaccine and other vaccine containing 17D strain, raised in chicken embryos.
Kyasanur forest disease (KFD)	*Flavivirus*	India	Tick (*Haemaphysalis spinigera*) bite.	Fever, headache, hemorrhages on mucous membrane *etc.*	Control of ticks. Use of protective clothing. Vaccination of people with killed KFD vaccine in endemic areas.
Omsk hemorrhagic fever (OHF)	*Flavivirus*	Russia	Ticks (*Dermacentor reticulatus* and *D. apronophorus*) bites.	Fever, head and body aches, vomiting, hemorrhagic diathesis *etc.*	Control of ticks. Use of protective clothing. Vaccination with CEE or RSSE virus vaccine (protective effect for OHF).
Tick-borne encephalitis (CEE and RSSE).	*Flavivirus*	Worldwide	Ticks (*Dermacentor, Ixodes* and *Haemaphysalis* spp.) bites.	Meningoencephalitis with biphasic clinical course *etc.*	Control of ticks. Use of protective clothing while visiting the endemic forest areas. Pre-exposure prophylaxis with Anti-TBE hyperimmunoglobulin.

Disease	Causative Agent	Geographic Occurrence	Transmission	Characteristics in Humans	Management in Humans
Zoonoses caused by Bunyaviruses: Crimean-Congo hemorrhagic fever (CCHF), *Hantavirus* infections and Rift valley fever (RVF)					
Crimean-Congo hemorrhagic fever	*Nairovirus*	Asia and Africa	Ticks (*Hyalomma* and *Ixodes* spp.) bites.	Fever, hemorrhagic diathesis *etc.*	Control of ticks. Use of protective clothing. Vaccination with inactivated mouse brain vaccine.
Hantavirus infections	*Hantavirus*	Asia and Europe	Contact of water contaminated with mouse excrements. Inhalation of aerosols contaminated with mouse excrements.	Fever, erythema, respiratory distress, kidney dysfunction *etc.*	Control of mouse population. Personal hygiene. Use of face mask by laboratory personnel while handling infected specimens.
Rift Valley fever	*Phlebovirus*	Africa	Contact with diseased animals during slaughter or parturition. Mosquitoes (*Aedes* and *Culex* spp.) bites.	Fever, rigor, malaise, maculopapular rash *etc.*	Control of mosquitoes. Use of alpha interferon, hyper-immune globulin and ribavirin. Vaccination of domestic animals with formalin inactivated and attenuated live virus vaccine.
Zoonoses caused by Filoviruses: *Ebola* virus hemorrhagic fever and *Marburg* virus hemorrhagic fever					
Ebola virus hemorrhagic fever	*Ebola* virus	Africa	Contact	Fever, hemorrhages *etc.*	Avoid direct contact with infected persons. Strict hygienic measures. Safe handling of infected persons as well as infected materials. Perform experimental work in BSL-4 laboratories. Control of hemorrhagic diathesis by fresh-blood transfusions (strictly avoid exchange transfusion).
Marburg virus hemorrhagic fever	*Marburg* virus	Africa and Europe	Contact with infected monkey blood or organs. Inhalation of infected aerosols.	Fever, maculopapular rash, nausea, vomiting, diarrhea *etc.*	Strict hygienic measures. Avoid direct contact with infected persons. Avoid transportation of patient in the acute phase. Use of high pressure suits, gloves, goggles and face masks by hospital staff.

Fungal zoonoses

Disease	Causative Agent	Geographic Occurrence	Transmission	Characteristics in Humans	Management in Humans
Aspergillosis	*Aspergillus fumigatus, A. flavus, A. nidulans* and *A. niger.*	Worldwide	Inhalation of air contaminated with spores of the causative agent.	Invasive aspergillosis, paranasal granuloma, aspergilloma of lungs *etc.*	Hygienic measures. Health education to the public. Treatment with corticosteroids, amphotericin B and surgical excision of affected part.
Blastomycosis	*Blastomyces dermatitidis*	America and Africa	Inhalation of air contaminated with spores of the causative agent.	Abscess and granulomatous lesions in the lungs, bone and other organs, fever, cough, chest pain, ulceration on skin *etc.*	Avoid contamination of open wound. Disinfection with formalin in the endemic areas. Treatment with amphotericin B.
Candidosis	*Candida albicans* and *C. tropicalis*	Worldwide	Contact with infected tissues.	Development of discrete patches on the mucous membrane of mouth, vagina, penis *etc.*	Treatment with ketoconazole and amphotericin B.
Coccidioido-mycosis	*Coccidioides immitis*	America	Inhalation of air contaminated with spores of the organism.	Chronic cavitating pulmonary lesions, fever, cough, chest pain *etc.*	Avoid exposure to contaminated soil or dust. Prevention of exposure of wound while handling the infected animals. Treatment with amphotericin B.
Cryptococcosis	*Cryptococcus neoformans*	Worldwide	Inhalation of contaminated air.	Development of nodules in the lungs, meningoencephalitis *etc.*	Avoid exposure to contaminated dust or soil. Treatment with amphotericin B.
Dermato-phytosis	*Microsporum* and *Trichophyton* spp.	Worldwide	Contact with infected animals.	Hair break, scales, crusts, inflammation of hair follicle *etc.*	Follow the strict hygienic measures. Treatment with ketoconazole or griseofulvin.
Histoplasmosis	*Histoplasma capsulatum*	Worldwide	Inhalation contaminated dusts.	Development of large cavities in the lungs.	Disinfection of soil with formalin in endemic areas. Avoid exposure to contaminated niche. Treatment with amphotericin B.

Disease	Causative Agent	Geographic Occurrence	Transmission	Characteristics in Humans	Management in Humans
Rhinos-poridiosis	*Rhinosporidium seeberi*	Asia and America	Contact with contaminated soil, dust and stagnant water.	Development of large polyps or wart like lesions in the nose or conjunctiva *etc.*	Avoid exposure to contaminated soil or dust and stagnant water. Treatment by surgical excision of the polyps.
Sporotrichosis	*Sporothrix schenckii*	Worldwide	Cat scratches and bites.	Nodules on inoculation site, mouth, nose, trachea *etc.*	Avoid contact with rotted wood, plant materials and infected animals. Treatment with potassium iodide solution or amphotericin B.

Parasitic zoonoses

Zoonoses caused by protozoa: Amebiasis, Babesiosis, Balantidiasis, Cryptosporidiosis, Giardiasis, Sarcosporidiosis, Toxoplasmosis, African trypanosomiasis, American trypanosomiasis, New World cutaneous leishmaniasis, Old World cutaneous leishmaniasis and Visceral leishmaniasis.

Disease	Causative Agent	Geographic Occurrence	Transmission	Characteristics in Humans	Management in Humans
Amebiasis	*Entamoeba histolytica*	Worldwide	Ingestion of food and water contaminated with cysts of the parasite.	Dysentery, abscess in liver and other organs *etc.*	Strict hygienic measures. Prevention of contamination of food and water. Treatment with metronidazole or tinidazole and paromomycin.
Babesiosis	*Babesia* spp.	Worldwide	Ticks (*Dermacentor, Hyalomma, Ixodes* and *Rhipicephalus*) bites.	Fever, hemolytic anemia, hemolytic haemoglobinuria *etc.*	Control of ticks. Use of protective clothing. Vaccination of dogs with antibabesial vaccine. Treatment with clindamycin and quinine sulfate.
Balantidiasis	*Balantidium coli*	Worldwide	Ingestion of food contaminated with pig faeces (cysts).	Intermittent diarrhea, abdominal pain *etc.*	Strict hygienic measures. Prevention of contamination of food and water. Treatment with metronidazole.
Cryptos-poridiosis	*Cryptosporidium parvum*	Worldwide	Ingestion of food and water contaminated with faeces of infected animals (oocysts).	Profuse watery diarrhea, cholengitis, pancreatitis *etc.*	Disinfection of drinking water. Avoid direct skin contact with human or animal faeces or garden soil. Avoid swimming in public bath. Treatment with paromomycin plus azithromycin.
Giardiasis (Lambliasis)	*Giardia intestinalis* (also known as *G. lamblia* or *G. duodenalis*)	Worldwide	Ingestion of food and water contaminated with faeces of infected animals (cysts).	Diarrhea, flatulence, abdominal cramps *etc.*	Strict hygienic measures. Prevention of contamination of food and water. Treatment with metronidazole, tinidazole, furazolidone or paromomycin.

Disease	Causative Agent	Geographic Occurrence	Transmission	Characteristics in Humans	Management in Humans
Sarcosporidiosis	*Sarcocystis bovihominis* (also known as *S. hominis*) and *S. suihominis*	Worldwide	Ingestion of infected beef or pork (cysts).	Diarrhea, dizziness, cysts in cardiac and skeletal muscles *etc.*	Adequate heat treatment of beef and pork. Avoid use of human faeces as fertilizer for pasture grounds.
Toxoplasmosis	*Toxoplasma gondii*	Worldwide	Ingestion of infected meat (cysts or sporulated cysts).	Fever, encephalopathy, lymphadenopathy, chorioretinitis, abortion, stillbirth *etc.*	Prevention of cats from hunting of potential intermediate hosts (mice or rats) or access of raw meat or slaughter wastes. Careful handling of cat faeces. Adequate heat treatment of pork and mutton. Treatment with pyrimethamine plus sulfadiazine.
African trypanosomiasis (Sleeping sickness)	*Trypanosoma brucei gambiense* and *T. brucei rhodesiense*	Africa and Asia	Tsetse flies (*Glossina palpalis* and *G. morsitans*) bites.	Lesion on inoculation site, splenomegaly, anemia, generalized lymphadenopathy *etc.*	Control of tsetse flies. Use of protective clothing and insect repellents. Avoid visiting of endemic areas. Treatment with suramin, pentamidine isethionate or melarsoprol.
American trypanosomiasis (Chagas' disease)	*Trypanosoma cruzi*	America	Bloodsucking bugs (*Triatoma* spp.) bites.	Erythematous swelling with pain at site of entry, regional lymphadenopathy *etc.*	Control of triatomine bugs. Use of insect repellents in bug-infested endemic areas. Treatment with nifurtimox and benznidazole.
New World cutaneous leishmaniasis (American leishmaniasis)	*Leishmania mexicana, L. braziliensis complexes* and other species.	America	Sandflies (*Lutzomyia* spp.) bites.	Similar cutaneous lesions as Old World cutaneous leishmaniasis.	Control of sandflies. Treatment with pentavalent antimony compound such as sodium stibogluconate.
Old World cutaneous leishmaniasis (Oriental sore, Delhi boil, Aleppo)	*Leishmania tropica minor* and *L. tropica major*	Asia and Africa	Sandflies (*Phlebotomus sergenti* and *P. papatasi*) bites.	Reddish-blue papule with ulceration, multiple cutaneous nodules *etc.*	Control of sandflies. Use of sandfly repellents (hexamethylbenzamide or diethyl toluamide). Treatment with sodium stibogluconate or meglumine antimonite.

Disease	Causative Agent	Geographic Occurrence	Transmission	Characteristics in Humans	Management in Humans
Visceral leishmaniasis (Kala-azar or black disease)	*Leishmania donovani donovani, L. donovani chagasi* and *L. donovani infantum.*	Asia, Africa and America	Sandflies (*Phlebotomus argentipes, P.papatasi, Lutzomyia intermedius* and *L. longipalpis*) bites.	Fever, itching papule with erythematous rim, anemia, hepatomegaly, lymphadenopathy etc.	Avoid contact with sandflies. Use of sandfly repellents (hexamethylbenzamide or diethyl toluamide). Treatment with sodium stibogluconate.

Zoonoses caused by trematodes: Cercarial dermatitis, Clonorchiasis, Dicrocoeliasis, Fascioliasis, Fasciolopsiasis, Opisthorchiasis, Paragonimiasis and Schistosomiasis.

Disease	Causative Agent	Geographic Occurrence	Transmission	Characteristics in Humans	Management in Humans
Cercarial dermatitis	*Austrobilharzia, Trichobilharzia, Gigantobilharzia* and *Schisto-somaticum*	Worldwide	Contact with cercariae liberated in water from snail (intermediate host).	Redness of skin, itching, malaise etc.	Avoid bathing or swimming in lake and coastal areas infested with the parasite. Keep away the ducks or other waterfowl from lake infested with parasite. Control of snail population.
Clonorchiasis	*Clonorchis sinensis*	Asia	Ingestion of freshwater fish (second intermediate host) containing metacercariae. Snail is the first intermediate host.	Fever, hepatomegaly, icterus, diarrhea etc.	Adequate cooking or frying of fish. Avoid introduction of human faeces into fish waters. Control of snail population. Treatment with Praziquantel or albendazole.
Dicrocoeliasis (Distomatosis)	*Dicrocoelium dendriticum*	Worldwide	Ingestion of ant (second intermediate host) containing metacercariae in water or adhering to fruits of water plants. Snail is the first intermediate host.	Cholangitis, hepatomegaly, icterus, splenomegaly etc.	Carefully watch for ants if eating fallen fruits from natural habitat. Avoid chew on grass blades of plants growing in water. Treatment with praziquantel.
Fascioliasis	*Fasciola hepatica* and *F. gigantica*	Worldwide	Ingestion of metacercariae adhering to plants and their fruits. Snail is the intermediate host.	Fever, anemia, peritonitis, fibrosis and calcification of bile ducts, icterus etc.	Avoid chew on grass blades. Avoid consumption of watercress from natural habitat. Avoid consumption of unboiled or unfiltered surface water. Treatment with triclabendazole.
Fasciolopsiasis	*Fasciolopsis buski*	Asia	Ingestion of metacercariae adhering to water plant and its fruits. Snail is the intermediate host.	Diarrhea, gastric pain, ascites, anemia etc.	Avoid consumption of raw water nuts and chestnuts. Prevention of contamination of water with pig faeces. Control of snail population. Treatment with praziquantel.

Disease	Causative Agent	Geographic Occurrence	Transmission	Characteristics in Humans	Management in Humans
Opisthorchiasis	*Opisthorchis felineus* and *O. viverrini*	Asia and Europe	Ingestion of freshwater fish (second intermediate host) containing metacercariae. Snail is the first intermediate host.	Cholangitis, liver cirrhosis *etc.*	Proper cooking of fish before their consumption. Control of snail population. Treatment with praziquantel.
Paragonimiasis (Pulmonary distomatosis)	*Paragonimus westermani*	Asia and America	Ingestion of meat of freshwater crabs and other crustaceans (second intermediate host) containing cercariae. Snail is the first intermediate host.	Peritonitis, pleurisy, pneumonia, chest pain, fever *etc.*	Avoid consumption of raw crustaceans. Control of snail population. Treatment with praziquantel.
Schisto-somiasis (Bilharziosis)	*Schistosoma mansoni, S. haematobium* and *S. japonicum*	Asia, Africa and America	Contact of skin with cercariae contaminated water. Snail is the intermediate host.	Pruritus, fever, headache, uropathy, hepatomegaly, diarrhea *etc.*	Avoid contact of surface water in endemic areas. Prevention of contamination of water with human faeces and urine. Control of snail population. Treatment with praziquantel.
Zoonoses caused by cestodes: Coenurosis, Diphyllobothriasis, Dipylidiosis, Echinococcosis, Hymenolepiasis, Sparganosis and Taeniasis.					
Coenurosis	*Taenia multiceps*	Worldwide	Ingestion of vegetables contaminated with eggs from canines faeces. Human and sheep are intermediate hosts. Canines are the final host.	Meningoencephalitis, headache, nausea, vomiting, dizziness *etc.*	Proper washing of vegetables. Prevention of dogs to access slaughterhouse wastes. Deworming of dogs. Treatment with praziquantel. Surgical removal of coenuri.
Diphyllo-bothriasis	*Diphyllobothrium latum*	Worldwide	Ingestion of freshwater fish (second intermediate host) containing plerocercoid. Copepods (*Cyclops* spp.) are the first intermediate host.	Megaloblastic anemia, neurological disorders *etc.*	Proper cooking of fish before its consumption. Treatment with niclosamide, praziquantel or mebendazole.
Dipylidiosis	*Dipylidium caninum*	Worldwide	Ingestion of cysticercoids or infected fleas. Flea is the intermediate host. Dog and cats are the final hosts.	Bloody diarrhea, abdominal discomfort, pruritus ani *etc.*	Control of fleas. Avoid licking of hands by dogs or cats. Treatment with niclosamide, praziquantel or mebendazole.

Disease	Causative Agent	Geographic Occurrence	Transmission	Characteristics in Humans	Management in Humans
Echinococcosis	*Echinococcus granulosus, E. multilocularis, E. vogeli* and *E. oligarthus.*	Asia, America and Europe	Ingestion of food and water contaminated with eggs from faeces of the final host (dog and fox). Inhalation of contaminated dusts. Herbivores (sheep, goat, and cattle) are the intermediate host.	Cysts in liver, lungs and other organs, icterus, hepatomegaly *etc.*	Avoid consumption of raw meat. Prevention of dogs to access slaughterhouse wastes. Surgical removal of cysts. Treatment with albendazole.
Hymenolepiasis	*Hymenolepis nana*	Worldwide	Ingestion of eggs or infected flour beetles (intermediate hosts). Human-to-human transmission occurs but does not require intermediate host.	Diarrhea, abdominal pain *etc.*	Keep the foods away from rodents and insects. Follow the strict hygienic measures. Treatment with praziquantel.
Sparganosis	*Spirometra* spp.	Worldwide	Ingestion of fish, frogs and snails (second intermediate host). Copepods are the first intermediate host.	Subcutaneous painful nodules, ophthalmodynia, seizures *etc.*	Prevention of contamination of water. Adequate cooking of meat of fish, frogs and snails. Surgical removal of larvae.
Taeniasis	*Taenia saginata* and *T. solium.*	Worldwide	Consumption of infected beef and pork.	Diarrhea, cysticercosis *etc.*	Adequate cooking of beef and pork. Treatment with niclosamide or praziquantel.

Zoonoses caused by nematodes: Angiostrongyliasis, Anisakiasis, Filariasis, Larva migrans visceralis, Oesophagostomiasis, Strongyloidiasis, Thelaziasis and Trichinosis.

Disease	Causative Agent	Geographic Occurrence	Transmission	Characteristics in Humans	Management in Humans
Angio-strongyliasis	*Angiostrongylus cantonensis*	Asia and America	Ingestion of crabs, crayfish and freshwater fish (paratenic host). Snail is an intermediate host. Rats are final host.	Eosinophilic meningoencephalitis, diarrhea, fever, severe headache *etc.*	Avoid consumption of raw crabs, prawns and crayfish. Boiling of surface water before its drinking. Treatment with mebendazole.
Anisakiasis (Herring worm disease)	*Anisakis simplex* and *Pseudo-terranova* spp.	Asia and America	Ingestion of saltwater fish and squids. Crustaceans are the intermediate host. Whales, dolphins and seals are the final host.	Eosinophilic granuloma and ulcers in the intestinal tract.	Avoid consumption of raw saltwater fish of endemic areas. Treatment with mebendazole.

Disease	Causative Agent	Geographic Occurrence	Transmission	Characteristics in Humans	Management in Humans
Brugia filariasis	*Brugia malayi* and *B. timori*	Asia, Africa and America	Mosquitoes (*Aedes* spp. and *Mansonia* spp.) bites.	Nodules in skin and cornea, eosinophilic granulomas of lungs *etc.*	Control of mosquitoes. Treatment with diethylcarbamazine. Surgical removal of nodules and parasites.
Dirofilariasis	*Dirofilaria immitis* and other *Dirofileria* spp.	Worldwide	Infected mosquitoes and simulia (blackflies) bites.	Chest pain, coughing, hemoptysis, localized vasculitis, pulmonary infarcts, granuloma around the worms, nodules in the eyes causing pain and itching *etc.*	Use of insect repellents. Protection from biting insects. Treatment by surgical excision of parasites.
Larva migrans cutanea (cutaneous larva migrans or creeping eruption)	*Ancylostoma braziliense* and *Uncineria stenocephala.*	Worldwide	Contact of bare foot with soil contaminated with infected dog faeces (larvae).	Papule at the penetration site, appearance of burrows on skin surface, itching *etc.*	Avoid barefoot walking on ground. Prevention of ground from contamination with dog and cat faeces. Treatment with albendazole or ivermectin.
Larva migrans visceralis (visceral larva migrans)	*Toxocara canis* and *T. mystax*	Worldwide	Ingestion of food and water contaminated with embryonated eggs from infected dog faeces.	Fever, abdominal pain, hepatomegaly, seizures, ocular larva migrans *etc.*	Avoid playing of children in contaminated ground. Treatment with albendazole. Removal of ocular larvae by surgery or photocoagulation.
Oesophago-stomiasis	*Oesophago-stomum bifurcum, O. stephano-stomum and O. aculeatum*	Africa	Ingestion of vegetables contaminated with third-stage larvae adhering on vegetables.	Nodules in the intestinal wall, obstruction and pain in intestine *etc.*	Follow the strict personal hygiene. Treatment with albendazole. Surgical resection in severe alteration of the intestine.

Disease	Causative Agent	Geographic Occurrence	Transmission	Characteristics in Humans	Management in Humans
Strongyloidiasis	*Strongyloides stercoralis* and *S. fuelleborni.*	Worldwide	Percutaneous infiltration of third-stage larvae.	Nausea, diarrhea, abdominal pain, urticaria etc.	Avoid barefoot walking in endemic areas. Treatment with albendazole.
Thelaziasis	*Thelazia callipaeda* and *T. californiensis*	Asia and US	Licking of conjunctival fluid by flies (*Musca* ssp. and *Fannia* spp.).	Conjunctivitis.	Removal of worms from eyes.
Trichinosis (Trichinellosis)	*Trichinella spiralis* and *T. pseudospiralis*	Worldwide	Ingestion of raw meat.	Fever, nausea, epigastric pain, vomiting, diarrhea etc.	Prevention of carnivores from accessing the raw meat in endemic areas. Adequate cooking or freezing (-25°C or below) of meat. Treatment with albendazole.
Myiasis	Dermal myiasis (*Cordylobia anthropophaga* and others). Wound myiasis (*Lucilia* spp. and *Caliphora* spp.).	Worldwide	Deposition of eggs by Diptra female flies at the margin of dermal lesions.	Swelling, painful skin lesions etc.	Proper disposal of food residues and garbage. Removal of larvae. Treatment with antibiotics.
Tungiasis	Sandflea (*Tunga penetrans*)	Worldwide	Invasion of skin by adult female sandflea.	Pruritus, inflammation, lymphangitis, painful skin lesions etc.	Hygienic measures. Application of topical antibiotics on the lesions.

Index

C

www.ingramcontent.com/pod-product-compliance
Lightning Source LLC
Chambersburg PA
CBHW060249230326
41458CB00094B/1625